Valley Animals

Valley Animals

True Stories About the Animals and People

of California's Santa Ynez Valley

•

Brooks Firestone

Illustrations by

Alasdair Hilleary, a.k.a. Loon

FITHIAN PRESS, MCKINLEYVILLE, CALIFORNIA, 2010

To my wife, Kate Firestone,
who loves and supports Valley animals
and her husband

The interior design and the cover design of this book are intended for and
limited to the publisher's first print edition of the book and related market-
ing display purposes. All other use of those designs without the publisher's
permission is prohibited.

Published by Fithian Press
A division of Daniel and Daniel, Publishers, Inc.
Post Office Box 2790
McKinleyville, CA 95519
www.danielpublishing.com

Distributed by SCB Distributors (800) 729-6423

Book design: Eric Larson, Studio E Books, Santa Barbara

LIBRARY OF CONGRESS CATALOGING-IN-PUBLICATION DATA
Firestone, Brooks.
 Valley animals : true stories about the animals and people of California's Santa
Ynez Valley / by Brooks Firestone.
    p. cm.
 ISBN 978-1-57464-500-2 (pbk. : alk. paper)
1. Animals—California—Santa Ynez River Valley—Anecdotes. 2. Domestic
animals—California—Santa Ynez River Valley—Anecdotes. 3. Santa Ynez
River Valley (Calif.)—Anecdotes. I. Title.
 QL164.F56 2010
 636.009794′91—dc22
                    2010013355

Contents

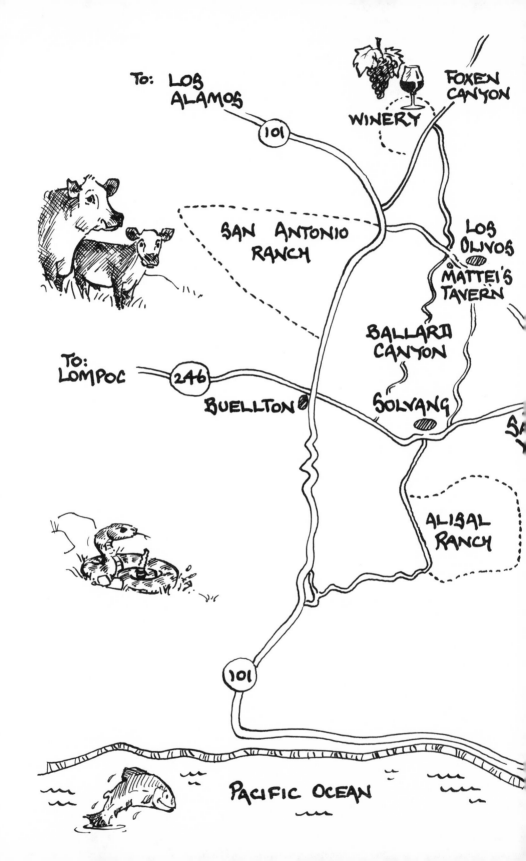

LOS PADRES
NATIONAL FOREST

TA
EZ

154

LAKE
CACHUMA

TO:
SAN
MARCOS
PASS,

AND
SANTA
BARBARA

Introduction

THIS IS a book about animals and people who live in California's Santa Ynez Valley. The past and present relationship of Valley people with animals, and the things that our local animals get up to, is an ongoing source of many fascinating stories.

All the stories included in this book are true. Sometimes the details might have become a little fuzzy over time, but each vignette is based on a factual event. There are alterations to protect the source or marginally legal interaction of individuals and animals, but the basics are factual. As the saying goes, you couldn't make up stuff like some of these stories.

In 1972, our family, Kate and I with Hayley, twelve, Adam, eleven, and Polly, eight, were lucky enough to move to the Santa Ynez Valley from our former home in London. In 1975, Andrew was born into the family. Life in our new home mostly involved growing grapes and starting up a winery, and this new career demanded most of our working time. But we also lived on and farmed a 2300-acre cattle ranch, the San Antonio, three miles north of Buellton on Highway 101. For twenty-five years, before moving to the winery property in the mid-nineties, we experienced all the country events a family would encounter growing up in the Valley and managing a ranch property.

Valley ranching inevitably involves many animal adventures of one sort or another. Fortunately, from time to time I made notes of our own animal moments as well as the stories we heard in passing. I have kept up these observations now for over thirty-five years of Valley life and, more recently, have sought out and researched more stories for this book.

The Santa Ynez Valley is forty-five minutes north of Santa Barbara, and about two and half hours from Los Angeles. Some twenty-five thousand people share the towns of Santa Ynez, Solvang, Buellton, Los Olivos, and surrounding ranchland. The Santa Ynez River runs through the center, and we are roughly twenty miles by ten miles in an irregular shape, surrounded by mountains. U.S. Highway 101 cuts through one edge of the Valley. There are some important people, some modest people, some characters, and a lot of pleasant, normal types living here. Solvang is a thriving Danish community and the tourist industry is important to the economy. Also, there are many retired residents and many hardy farming and ranching types. In recent years, vineyards and wineries have become an important part of the Valley.

Household animals, wild animals, commercial animals, working animals, and competitive animals all exist side by side with humans, and the relationship between the two is a source of ongoing fascination. Valley people tend to be creature-friendly and involved with animals. We are country people in that respect, and people living here are often judged by their relationship, treatment, and understanding of animals. What the animals think of us, we will never know, and sometimes there is a suspicion that they are as much in charge of events as people; certainly equally interesting.

The content of animal stories in this book is of necessity a chance and random selection because these are the ones I have experienced or heard of. There are countless other dramatic stories that are not included here simply because I have not heard them to tell. Forgive this book the missing sagas that did not make it to this collection and could probably fill a few more books.

The primary motivation in my telling these stories is that through them we might better understand the culture, traditions, and philosophies of the Valley. There is a generosity of spirit here, but also a down-to-earth attitude that makes this a real place. Animals help us keep our sense of proportion and better nature. In pondering these stories, we can learn much of who we are.

I owe a debt of gratitude to all who have sat down and told their stories and shared their animal adventures. Without the memory and retelling by countless Valley people, these stories would never be written here, and I am very thankful for everyone's help and willingness to relate their experiences. Thank you also to Jim Lindsey, who helped me edit. And mostly, thanks to my understanding wife, Kate, who put up with my passion to tell these sagas and who also helped edit.

My fondest hope is that, in contemplating ourselves through past and present animal events, we will more completely understand our Valley and better preserve this very pleasant land.

Part One

Wild Things

Good Samaritans

FROM SOUTHERN California the entrance to the Valley is over the beautiful San Marcos Pass. Most Valley people take a deep breath and know they are home when they reach this summit from Santa Barbara and see the rolling hills ahead. In the late 1960s, two Los Angeles traveling businessmen were driving that route. They were city types and not locals familiar with our animal friends.

As they started down the grade into the Valley, they saw a dog hit by a glancing blow from a car and left lying at the side of the road. The travelers stopped to see the dog was breathing but unconscious and were moved by their dog–man compassion to save the hurt animal if they could. They loaded the unconscious stray into the back seat of the car and determined to drop him off at the nearest town veterinarian until he could mend and the owners could be found to reclaim the dog. They were filled with righteous, warm intentions.

About halfway to Los Olivos, a few miles down Highway 154, the dog in back suddenly came to life and awoke with a snarling, ferocious menace. An incredible, threatening growl roared from the back seat and the dog crouched with bared fangs ready to kill. The Los Angeles men recognized a serious threat and were up to the challenge. The passenger reached down for a handy flashlight, fortunately in the car side pocket, and decisively cold-cocked the dog, who sank back in the seat, unconscious for the second time.

They debated leaving the animal at the side of the road again, but, armed with a ready flashlight club, the two drove on to Los Olivos. Here, they soon found Dr. Ayre's veterinary clinic, where Judy Hale's art store now stands, in the center of town at Alamo Pintado and Grand Avenue. The men brought the sympathetic vet out to the car and asked if the clinic could take the animal in and find the owners.

Dr. Ayre took one look and exclaimed to the astonished men, "That's no dog, that's a coyote!"

This story was still making the rounds in the early '70s and is one of the inspirations for this book. In this incident is the urban as opposed to Valley understanding of our animal population. Of course when the coyote woke up after being dumped back in the wild, he probably had serious reservations about the people population and doubtless stayed away from roads.

· 𝒥 ·

Overnighters

ONE FALL morning in the early '70s, Kate and I woke before dawn and decided to go for an early ride just for fun. We left the children sleeping and saddled two patient horse friends by the tack room light.

In the early '70s, our ranch, before we bought the big ranch north of Buellton, was about a mile from Los Olivos toward the mountains. We decided to ride toward town and traveled slowly in the predawn shadows from our ranch down Calkins Road toward a very quiet little Los Olivos. An unusual escapade for the two of us, but it seemed right at the time. In those years, there were no homes on Calkins Road and the morning was quiet, dark, and deserted.

There was a crystal-clear chill in the air and a stillness in the fall predawn as the rising light brought shapes to the hills and trees. An eighty-acre field across a creek bed had been dry-farmed with barley that year and harvested, leaving a yeasty, fresh, country smell. The barley leftovers must have served as inviting fare for a crowd of amazing visitors.

As we silently rode down the side of the road we sensed a movement in the barley field about two hundred yards across the creek bed. There was an unusual stirring movement, somewhat sinister in the half light. A soft commotion of dark, lumpy things heaved and simmered, sending shivers of the unknown up our spines as we quietly asked each other what in the world was going on over there.

We stopped to watch, and, if the truth be told, stood ready to run back to the ranch if a threat materialized. We peered into the growing light. The minutes ticked by until dawn revealed the most amazing and spectacular animal sight we will ever see in the Valley.

At least five hundred, and maybe many more, Canadian Honker Geese were stirring. They stumbled about on their web feet, then ran and pumped themselves into the air with a wing woofing sound. The birds rose against the red dawn and distant blue sky with a circling grace above us, then silently mustered into a formation that finally swept off to the south. The moments of this inspirational sight passed too soon.

I understand from other ranchers that the sight of that many overnight Canadian Honkers on their annual migration south is a rare sight in the Valley, and it was a total surprise to us newcomers. The geese were majestic, spiritual, and purposeful with a unique power and grace in the early light.

We gaped and dried our eyes as they sailed away, realizing that we had been treated to a near miracle of beautiful animal encounter.

· *J* ·

Deputy Dogs

RON CARLENTINE is a local who has ranched in the Valley and now works in property management for Santa Barbara County. In the '70s, he was manager of the El Capitan Ranch and Campground on the coast, off Highway 101. One of his more pleasant responsibilities was maintaining a fenced animal petting area for the enjoyment and learning experience of young camp visitors.

The area contained chickens, lambs, rabbits, emus, young goats, and an aged ram that occasionally and harmlessly butted the visiting youngsters. Children would take in twenty-five-cent cups of grain and feed the animals, who thought the kids were great and performed the role of contented recipients of child curiosity and affection quite happily. This peaceful scene was brutally destroyed one night by a rogue mountain lion.

Ron arrived in the early morning, the day after the crime, to clean the area and made the heartrending discovery that every animal had been ravaged and destroyed, with the brutalized remains scattered around the petting area. A young goat had vanished, probably carried off to provide a couple days' food for the killer, but the other animals were victims of wanton brutality. Ron quickly fenced off that area of the ranch to prevent children from seeing the mayhem.

In a rage at the unfair savagery, and also in the knowledge that a rogue lion might come again to the camp, Ron called a friend, the central California Federal Trapper, who immediately responded. These Federal employees protect and manage national lands, where the lion had come from, and are salty characters, wise to the ways of animals. The friend surveyed the scene of destruction and said, "This is a job for my partner in Sacramento."

That afternoon, the northern California Federal Trapper arrived by plane. Airport employees carefully unloaded six cages from the luggage compartment. Hearty hunting dogs will put up with almost anything in the name of sport, and sometimes country hunters can be seen driving with pickups stacked with cages in the back of their trucks. These six Federal dogs had special light-weight containers and were accustomed to the call to duty that required sudden and long travel. As instructed, Ron had brought his pickup to the Santa Barbara airport and he directly drove dogs and trapper to the crime scene.

Most of us might imagine dogs like these as matched and handsome bloodhounds of the type pictured on the covers of mystery novels, or romantic fox-hunting hounds out of an English print. However, this pack of enforcers were an ill-matched and scruffy group of mixed-breed hard cases that lived to hunt. They were a pack in name only, but they worked as a team and, in their doggy way, knew their business very well. The equalizers had come to town with a vengeance.

On arrival at the camp, the trapper closely inspected the evidence for over an hour. Eventually he summed up the situation. "This is a mature female with one tooth missing who will kill again. We need to eliminate this threat and we'll find her first thing tomorrow." The dogs watched from their individual cages. They could smell the carnage and knew their job ahead. They had been there before.

The next day, in the early predawn darkness when the scent would be good, Ron, trapper, and pack arrived at the scene and swiftly and professionally swung into action. The keen dogs were let out of their cages and explored the petting area with quiet excitement. The trapper shouldered his rifle and silently signaled to the pack to find, motioning to the logical direction the killer cat would have taken. All was silent except the noise of dogs running through the brush. Ron and trapper walked behind the hunters, who dashed around the country in eager exploration.

Soon the quiet was broken by a dog baying the news that a scent had been found. At first a single voice, a tentative pause, and then a repeat broke the silence. Chills went down the backs of the two following men to hear the deep bay of a hound on the hunt. Almost immediately a second voice spoke and then a chorus confirmed the find. The baying dogs were now in full pursuit and the chase lasted for about fifteen minutes. The full, furious, and triumphant cries told the running men that the quarry was at bay, and they arrived, panting, to find the killer cat up a tree, growling and spitting at the surrounding pursuers.

The trapper dispensed justice with one shot and the big cat fell to the ground. The excited dogs circled and celebrated with triumphant voices and searches for approval. The trapper praised his pack and let them know of a job well done. Ron confirmed that it was, in fact, a large, mature female mountain lion with one large fang missing. That afternoon, he would bury the killer under the tree, knowing that a potent threat had been eliminated.

Trapper and dogs returned to the truck and cages. Later that day they all flew back to Sacramento to await another call. Even with the immediate threat eliminated, Ron could not bring himself to reinstitute another petting area. The Valley hills still hold mountain lions that do their lion thing, and, hopefully they will live in peace with us neighbors.

· 𝒥 ·

Bear Blockage

THE SAN Marcos Pass was dark and misty at five A.M. on a fall morning in 2006. Frank Troise, like many successful Valley entrepreneurs, sometimes found it necessary to drive to a Los Angeles office for his job. On this day, he departed very early from Los Olivos to beat traffic on the run to LA, congestion that so many of us in the Valley know too well. As he was tooling along in his Toyota Prius thinking about the day and, fortunately, not speeding in the fog, a large black bear straddling the road loomed directly in his headlights.

The good news was that Frank dodged the large, transfixed animal. The bad news was that the maneuver left him a hundred feet down the bank stuck in brush and trees, a passage that had totally wrecked his car but broken his fall. One of Frank's haunting memories of this incident is the sound of the foliage scraping against the sides and undercarriage of his little car as he made his plunge. He quickly took stock of his physical condition and thanked the Good Lord for no injuries and the three-way seat belt that had probably saved his life. Alone in the dark, he pulled out his cell phone and called 9-1-1.

The special Deputies that work emergency 9-1-1 sit at high-tech consoles in the Sheriff's headquarters north of Santa

Barbara. Their computers and phones are connected with Fire, Sheriff, Highway Patrol, Emergency Medical, with access to Federal, State, County and all other authorities and individuals that might be needed in any situation. Most of the shifts are quiet, but their jobs can suddenly become very interesting when quick judgment and a bank of resources are very necessary and life-saving. This weekday five A.M. call from a shaken up but intact motorist stuck on the San Marcos Pass was almost a welcome relief from the late-shift doldrums.

After assessing the situation in the controlled and well-modulated professional tones of the emergency dispatcher, Sheriff and Highway Patrol cars were sent to find the accident. The caller, Frank, was stuck, distressed, and uninjured, but otherwise in no threat and not blocking traffic. With no other 9-1-1 business and, in the line of normal duty, the dispatcher stayed on the line with Frank to reassure him and help in any way a distant voice could serve. The line was also open for the searching emergency people driving to help.

As time went on the deputies and Patrol could find no skid marks or other clues on the highway and were having trouble locating the stricken vehicle. The dispatcher talked to the searchers and Frank to try to help, but he seemed to be lost to the world. Eventually, and as Frank had no injuries and was not far from the road, the dispatcher suggested that he might simply exit the car and make his way to the road and flag down the authorities.

Frank agreed that he probably could manage, but did not want to. "Why not?" asked the dispatcher. "It would help in your rescue." "Well," Frank replied, *"Do you think I am getting out of this car with a bear outside my door?"*

9-1-1 had heard many stories, but this reply became a legend in their coffee breaks. Frank's exit from the highway was finally discovered and deputies and patrolmen, also conscious of bear threats, gave him an armed and well-lit path to safety. The bear was surely long gone and wandering.

· 𝓙 ·

Historical Birds

IT SEEMS like a vision from ancient history, but one of our Valley ranching families still remembers enjoying the company of condors during quiet evenings in the upper Happy Canyon. Condors were once a fixture in central California, but have become rare and endangered in modern times. The vulture-like birds are the largest in North America and soar high in the back country sky, seldom if ever seen now in the Valley.

Joe and Laura Alegria have leased the farming and ranching operation on the Phillips' Las Crusitas ranch since the early fifties. Joe is one of the great ranchers in our Valley and has a million Valley animal stories of his own, and one of his favorites was routine for him, but out of the question in our modern Valley.

Just up from his ranch home was a hill, grass-covered in the spring, but bald, brown, and eaten down in the summer months. All ranching operations include some casualties on the property, and if a cow died or Joe came on a deer or boar carcass he would drag the remains with his tractor to the top of that hill nearby his home. With the condor dinner set on the hill, Joe and Laura would sit on their porch in the evenings, relaxing after a day's work, and hope for some action.

They lived in condor territory in the fifties, and if they were watching on a lucky evening, they would spot the big birds circling high in the evening sky. The circles would diminish as the birds decelerated and lowered their altitude, with a keen bird eye on the hilltop, and soon the condors would swoop and awkwardly land. Sometimes the hill would entertain three or four of the beautiful and hungry visitors enjoying dinner on Joe. They would feed and relax for perhaps a half-hour and then the fun would begin when it was time to be airborne.

The big birds would gain a little momentum needed for flight by waddling down the hill, their stomachs now loaded with food, and their huge wingspan would barely lift them off the ground. Their flight path would then head straight for

Joe and Laura sitting on their porch, and the struggling condors would barely clear the Alegria home. Joe remembers the
swish of their wings as they flew low overhead. Soon with
growing momentum and a little altitude the ungainly birds
on land would again become incredibly graceful flying creatures. Once one bird left a twenty-five-inch feather as a sou-
venir that Joe kept for years.

Joe and Laura are in their nineties now, and they miss their
condor friends, with the scarcity and retreat of the big birds
into the Los Padres National Forest. But he remembers well
the beauty and rarity of a pleasant evening in their company.

· *Ꞙ* ·

Lion Laws

IN CALIFORNIA there are thought to be five thousand
adult mountain lions, more or less, although nobody knows
for sure. In 1972, strict laws were enacted to end hunting of lions, and in 1990, Proposition 117 declared the mountain lion
a "specially protected mammal" and further contained many
regulations for its preservation. Over the years, the growing
mountain lion population has led to the increased killing of
ranch animals and threats to people because, as many people
believe, the lions have lost all fear of humans. In 1996, growing lion problems led ranchers and victims to react with a ref-
erendum to repeal the protection laws. I was a member of the
California Assembly at that time and because Santa Barbara
is lion territory, I was drawn into the debate.

The Valley has a mountain lion population invisible to
all but the rare, accidental sighting. That is not to say that
many of us are not daily observed by lions going about their
territorial business. In twenty-five years of living on the San
Antonio cattle ranch north of Buellton I never saw a lion, although I did see many tracks and signs. Now and again I felt
observed in the back country with a little prickly sensation
up the neck, but I was never sure why.

Since 1990, there have been a number of California lion

attacks on humans and some fatalities. Our famous local attack occurred in 1992 when a father beat off a lion that had a
bite hold on his nine-year-old son. Fortunately, dad and chair
won the tussle and the boy lived with only a few scars. Lions
will attack women and children, and Valley hikers should
know this and be aware and preferably not travel alone. The
accepted protection is to stand still and try to look big. Maybe
a little prayer would help too.

One day a man was walking with a herd of goats on the
San Fernando Rey ranch when a lion dashed out of the bushes and carried off a young kid goat. I heard this story in the
company of Bill and Toni Simon, who had just bought the
ranch and were wondering about country ways. "What does
that mean?" asked Toni. "Well," I replied, "it means that you
should only walk on the ranch if you have the cover of a goat
herd!" She is a strong hiker and never took my advice.

A quarter of a mile from our San Antonio Ranch on Jonata
Road was a law enforcement pistol range where Sheriff's Deputies and Highway Patrolmen practiced and qualified. We
would hear the pop, pop of the pistols in the afternoons on a
regular basis. About 1990, a large sign was posted on the range
gate, right down our road, warning pistol shooters to "Beware,
Lions in Area!" Of course if armed deputies should be warned,
we were very concerned for the safety of five grandchildren
who then lived on the ranch, and the children were not allowed on the back of the property alone. I hated to give away
the freedom of the ranch from my grandchildren to the lions,
but that was the way it was.

The California Department of Fish and Game are in
charge of California animals, and take their lions very seriously. In their words, they have "The primary responsibility
for managing and protecting all fish and wildlife, including
their habitat." But certainly protecting the public and allevi-
ating damage to private property are equally important pri-
orities. I was happy to hear that my grandchildren were on
an equal protection level with the lions, but I knew if I ever
proactively tried to shoot a lion on the ranch I would be in big
trouble. Only if one of the lions actually killed or maimed a
child would the lion be in trouble also.

In the Valley, the pro-hunting and pro-protection debate proceeds, divided along discussion positions similar to the rest of California. The pro-hunting side maintains that lions are not endangered, but in fact increasing. They reference the livestock killings and the dangers to humans when lions proliferate and lose their fear of people. The protectionists long for a healthy environment for all creatures and reference the loss of our California symbol, the grizzly bear, as an example of what might happen to the lion. To them, the mountains are more interesting, beautiful, and natural with a healthy lion population, and animals must be preserved, whatever the inconvenience or threat. In the mid-nineties, the ranchers and hunters began collecting signatures for a state proposition designed to change the law and bring back hunting and control the growing lion population.

My part in the proposition campaign was to speak in support on the Sacramento Assembly floor and in legislative committees, and occasionally in public. My most important debate was in the Long Beach Auditorium where many interested people and a good deal of media would hear from the for and against sides in equal time slots. The anti proposition, lion protection, side were animal rights types who favored preservation of endangered species and opposed any animal hunting or harassment. The three speakers on our side were a professor who believed the lion population was not endangered and did not need protection, a rancher who would speak convincingly about the lion depredations of his herd in Santa Paula, and I would talk about what it meant to restrict grandchildren when deputies were afraid of lions in an area.

I thought we had a very good case, and all we wanted was some limited hunting to control the incursion of lions on ranches, but we had a problem. We were not able to control free speech, even on our own debate side, and a tough guy showed up who demanded his time to plead the rights of bowmen. Prior to 1990, people with bows and arrows were allowed to take out a permit and hunt lion with dogs or stalking, and that is what he demanded again.

We three scheduled speakers made our case with, I thought, dignity and credibility. We seemed to be winning the day and would get good press. We established that the lion population is growing rapidly and invading the suburbs, that lions ravage herds and savage little lambs, baby calves, and puppies, and that honest ranchers' grandchildren are hiding behind bolted doors. Then our bow-and-arrow friend got up to speak. I don't remember exactly what he said, but I can paraphrase his remarks as I am sure they were heard by the audience. He would have been translated as "We demand the right to go out with our four wheelers and beer coolers and turn our dogs loose on these furry, beautiful cats until we put them up a tree and shoot them full of arrows and they drop in agony." We lost that one. After much statewide effort and debates similar to our Long Beach experience, the proposition was defeated and the lions continue, protected from all harassment.

Meanwhile back in the Valley, the lions peacefully make their way in our hills unnoticed by most of us, and hopefully, this truce will remain.

· *J* ·

Bear Bathing

LAKE CACHUMA is a large body of water in the Valley that doubles as a County recreational park and source of water for the City of Santa Barbara. Visitors camp and fish, although swimming and high power boats are not allowed. It is a sparkling jewel surrounded by mountains, occasionally snow-capped in winter and flowered in spring.

Liz Gaspar is the Cachuma Lake Park Naturalist and is as good a host as any park could desire. Her duties include looking out for all the animals on, in, and surrounding Lake Cachuma as well as sharing her considerable knowledge with visitors. Occasionally she takes boat tours around the lake with those fortunate enough to be included.

One day, while she was shepherding a boatload of tourists, one of her visitors spotted an unusual object moving in the water and asked, "What's that black thing in the water?"

Liz steered her boat closer to discover a large bear swimming in the middle of the lake.

Bears can swim with a powerful dog paddle, as demonstrated by polar bears who cruise the arctic, but local Valley black bears usually don't, and this long trip across the lake was a mystery. Liz steered the boat behind the swimmer and soon realized that the bear was headed directly for the middle of a well-populated County campground. With no lake swimmers there would be no startling water encounter, but she realized that an emerging bear at picnic time could cause considerable consternation to the visitors, if not some potential threat.

Liz quickly got on the radio and alerted her fellow park rangers, who decided to evacuate families to high ground from the path the bear might take. The authorities and park visitors watched and awaited developments. The wooly intruder with snout and ears above the water paddled steadily toward the shoreline campground. Time seemed to stand still in the quiet afternoon as the bear made his steady progress. Liz and the boatload followed and all shore eyes, getting more country drama than usual, watched intently. The tension mounted.

Eventually the bear padded out of the water and shook a considerable shower of water off his coat. Bears do not see well, but this one raised his head and sensed, smelled, and peered at the audience of gaping visitors and officials in the distant welcoming party. He hesitated momentarily and then moved as rapidly as bears can for the far oak woods and wilderness. The park rangers were curious about where the animal might reappear and suspicious of a panic-stricken bear doubling back on the crowd. They jumped into their trucks and tried to follow, but this uninvited picnicker was long gone.

In telling this story, Liz explained that most animals are natural swimmers. She once followed a buck and two does swimming clear across the lake, and she has seen raccoons, bobcats and even rattlesnakes taking the occasional plunge. "But," I asked her, "why would a bear swim across the entire lake?"

Liz has handled tourists for years and looked at me with the wise eye of an experienced storyteller. "Well of course," she said, "to get to the other side!"

· *J* ·

Camp Mates

MIKE ALLEN, Santa Barbara County Clerk of the Board and outdoorsman, once set out with four friends to enjoy the distant Los Padres Forest back country on a camping trip. They back-packed on the trails to a remote campground near Painted Rock just off the Sierra Madre Ridge Road.

That evening, as they set down their gear near an old oak, they noticed an insistent and loud buzz from the tree, the sure sound of a large rattlesnake. They moved their gear to a picnic area further from the tree and sat down to enjoy a beer, packed in to celebrate their days off. One of the campers pointed in amazement and they all watched as a fat, ugly rattler, probably the same one that had warned them off the tree, emerged from a gopher hole near their beer fest. After seeming to survey the scene, the huge snake slithered off back to the protection of the oak. The unanimous decision of our campers was to move camp a little further off, assuring themselves the snake would probably stay put and they could rest more or less easy.

They cooked their first night's dinner and were joined by a lonely Fish and Wildlife condor specialist who would camp out for a few days at a time in a trailer set out on the ridge. His assignment was to observe, study, and assist the tenuous condor population whose territory was in the Los Padres back country. They visited at length about the condors and wilderness, and the condor man asked them to help him load a calf carcass into his truck. The carcass had been delivered to his storage trailer by the condor program to be delivered the next day to a remote feeding location. The campers helped pitch the rank carcass into the truck and returned to their sleeping bags for the night, down the trail from the truck and trailer.

The evening was quiet and beautiful on the remote hillside and they slept under the stars.

The next morning the Fish and Wildlife man had a harrowing story to tell. In the middle of the night he had heard noises from the vicinity of his truck and emerged from his trailer with a large flashlight. His light revealed a huge, huge bear rustling in the back of the pickup. To his amazement, he watched the bear swoop up the calf carcass in his paws and walk off with the entire dead animal carried in his jaws. The man followed the bear with his light, safely from the door of his trailer, as the bear, with the calf still in his jaws, sauntered off into the dark brush. The trailer walls seemed a little thin to the specialist for the rest of the night. Pondering the bear sighting, the campers hiked off in the midday to swim in the Sisquoc River, keeping a keen eye on the brush as they walked.

With snakes and bears playing on their imaginations, the campers had a watchful dinner back at their base camp. The giant bear never reappeared and the snake rested happily in the old oak. Just before turning in someone asked, "Anyone seen a lion today?" Valley animal mind games being what they are, their sleep that night was more watchful than restful.

· 𝒥 ·

Close Call

RATTLESNAKES ARE mean, ugly, and universally disliked by the Valley community for their potentially lethal ways. They get a worse rap than they deserve, because for the most part they would rather slither away from trouble than impose a confrontation, and their loud rattler buzz gives their location away in any case. Why creation put a deadly injection on one end and an alarm system on the other is one of the mysteries of the animal world, but anyone who has heard the dry, insistent angry rattle noise of an aroused snake will give the sound a wide berth.

Everyone in the Valley will kill a snake if they can with

stone or hoe or any weapon to hand. Once, rancher Dean Brown's guests were barred from the path to the house by an angry rattler and the snake was promptly shotgunned into oblivion. Another time an amplified drumming from under a slightly raised wooden dog bed caused the owner to stomp the wood, crushing the snake underneath and destroying the bed. Such is the life of rattlers who get more than they give with very few real snake bite tragedies. But they do happen.

Once a rattlesnake was picked up by a tomato harvesting machine that carries a line of workers sitting at a conveyor belt picking out the tomatoes as they are swept off the field and then up by the belt to be sorted. The snake traveled the belt along the line of workers, but the workers peeled off like a formation of fighter bombers going after an enemy carrier down below. The snake was carried the length of the belt and under the machine tires. When the incident settled down, there was no enthusiasm to be sitting first in the line.

Our family shepherd dog, Tucker, once went for a hike in the Los Padres Forest with son Andrew and his fiancée, Ivana. When they returned, Tucker did not look to be himself and on the edge of collapsing. Fearing a snakebite, we had him immediately in the car headed for Dr. Dean's clinic. On the way he deteriorated further, and I called ahead in panic. The sympathetic veterinarian met our car with an assistant and gurney to rush the failing dog into the treatment room. A half an hour and some powerful restorative medicines later, Tucker was alive, but stayed the night under close observation. He had been bitten on the nose by a rattler and was lucky to survive. Doubtless he had simply sniffed the snake to check it out and say hello, but rattlers are not sociable.

Country lore believes that rattlesnakes will seek water to wash after their winter hibernation, to clean off the snake-den smell and take on a fresh attitude. Whether that is true or not, snakes can and do swim, but most people do not look for a rattlesnake in the water.

The John Cody family lived for years in the back country of the Los Padres Forest, where Cody found rare stones to fashion into valuable sculptures. Valley people wish they had bought all the Cody work they could find in the early days before he had come into his own with an international reputation. Years ago, the Cody's son Ernie was twelve and full of a boy's curiosity as he was floating a raft on the Santa Ynez River on a lazy afternoon. He spotted a "dead" snake in the water and reached down and touched the animal. He was punished with a lightning strike to his finger, with two fangs injecting poisonous venom in a sudden and painful bite.

Ernie was a country boy and knew not to panic. He sat down on the bank, used his handkerchief for a tourniquet, and contemplated his swollen finger. His parents had left for the afternoon with the only car, but fortunately family friend David Davis was staying at the ranch and soon began to look for Ernie who had been out of sight too long. When he found Ernie he immediately sized up the situation and carried him to a beat-up truck and began the long trip to the Valley hospital. They soon ran low on gas, which forced a stop at a neighbor's ranch and delayed their long drive to town in the old truck. Ernie stayed cool and maintained the tourniquet but began to feel a little sick.

After two hours they finally arrived at the emergency room where the doctor on duty realized he had an urgent situation beyond the usual snake bite. He reached the emergency poison consultant at the University of Southern California, who advised a heavy-duty dose of anti-toxin. By now his anxious parents had caught up with their son as he fought for survival. He was momentarily better, but then had a near-fatal reaction to the medicine. Later, gangrene set into his right index finger where he was bitten, and he underwent a large dose of antibiotics. It was six days before he was well enough to be released from the hospital and back to the Las Padres home.

That incident ended well, thanks to a cool-headed boy, a quick-acting family friend, and a talented medical team, but the potential for tragedy is always there. John Cody found the snake, or at least one that looked like the one, and took out his protective father's revenge. Ernie never enjoyed the water quite the same afterwards.

· 𝓕 ·

Bobcat Nightmare

A GREAT DAY in the Valley is a peaceful boat trip on Lake Cachuma, and the best of all cruises would be guided by Liz Gaspar, Park Naturalist. She protects and studies all the animals in, around, and above Lake Cachuma and knows the habits and hideouts of most creatures who call the lake home. She is very accomplished in explaining the lake critters to visitors.

One day she was guiding a group of elderly tourists for a sunny day on the lake. A particularly sharp-eyed sightseer spotted a bobcat sound asleep in the grass beside the roots of an old oak tree. Liz cut the engine and drifted closer in the silence of the afternoon with the boatload of intently curious and hushed onlookers. For some unusual reason, the cat did not raise a watchful eye, but slept soundly in the afternoon quiet. The scene seemed too peaceful.

Soon, another tourist pointed to an approaching deer quietly moving toward the sleeping cat. The doe appeared alert and intent, stepping slowly and purposefully toward the comatose fur ball. Only the soft noise of a lake breeze in the trees and the occasional water ripple flavored the visitors' concentration. The boatload silently stared, the bobcat dozed, and the doe stalked in the sunny afternoon. This doe was no Bambi. Disneyland will portray dreamy-eyed, cuddlesome furry creatures who gambol on the flowered greenery, but nature is not quite like that idyllic scene. This Valley outing was about to have their innocent deer-creature illusions shattered by the harsh realities of wildlife.

Suddenly, the doe snorted and charged the now instantly awake and startled cat. In the short deer scramble to reach the victim, the cat vanished into the roots of the tree and disappeared. Obviously the tree was bobcat home and shelter, and also protection from the murderous deer. The frustrated, snorting, and pawing doe pranced in frustration while the tourists gaped. Their storybook notions of the peaceful, wide-eyed deer surrounded by butterflies and singing flowers were gone forever.

The fact is that a deer, and particularly a mother doe, can kill a bobcat in two seconds with thrashing legs and hoofs. A good friend, Jim Lindsey, while walking down his long driveway on Oak Trails Estates, actually witnessed two deer chase down a coyote and kill him on the run by dashing out his brains with their striking hooves. The deer in this case outnumbered the lone coyote, were faster, and sensed a powerful motivation of protecting the herd.

In the Lake Cachuma incident, the mother doe must have wandered by, with a young deer stashed somewhere back in the bushes, and determined to take out the potential bobcat threat, even though the cat was fast asleep under a tree. The fat and dozy bobcat never sensed the nightmare approaching, and was lucky to escape unharmed. Even on the doorstep of a safe house, that bobcat would always sleep less soundly in the future. The boatload continued with lost animal innocence.

· 𝕵 ·

Not Kittens

THE MYSTERIOUS appearances, the misunderstandings, and the relocation of visiting mountain lion cubs in Solvang will be the source of stories for many years.

The first sighting was by Jim Farnum when he saw and photographed three small mountain lions, probably cubs, on the Alisal golf course, a very unusual sight. The next day, Maren Thomas saw two lion cubs in her yard and had the extraordinary experience of watching one seem to want to play with her suspicious Yellow Labrador in the yard. The sightings and stories grew and people wondered, because we all know lions exist, but never on golf courses or in backyards.

The next day, Alisal resident Joan Speirs, when pulling into her driveway, was astonished to see two young cub lions walking across the lawn. She actually had the presence of mind to snap a photograph of the trespassers. Joan also had the sense to call her neighbors Steve and Lou Ann Bender, alerting them to the two young lions headed their way. Here, events took a tragic turn.

Steve Bender emerged from his back door to hear a growl and scuffle about thirty feet away in the bushes. He approached to see the larger lion cub emerge with their elderly cat, Stripes, in its mouth. Steve actually kicked at the cub, but then had second thoughts and retreated into his house. The Benders had the haunting experience of watching from the window while the two young lions consumed their pet cat before they departed.

The next incident took place the following day when the lions were spotted in Solvang on Third Street. Sheriff's deputies were quickly on the scene and soon cornered the two young lions against a fence. The situation was a standoff and weapons were drawn, not in fear of the young emaciated pair, who did not seem threatening, but in case their mother might be lurking and come to their defense. The Sheriffs immediately took the best course in these situations and called Julia Di Sieno, Director of the Valley Animal Rescue Team.

The Rescue Team is one of the great institutions in the Valley, dedicated to rescuing, rehabilitating, and releasing injured or orphaned wild animals and sometimes finding a home for domestic animals. Julia Di Sieno, a medical nurse in real life, looks more like a fashion model than an animal rescue person, and has run the Shelter for seventeen years. She has a passion and gift for her calling. Julia and her volunteers are always ready for an emergency and licensed by the California Department of Fish and Game to take in wild animals, but not mountain lions. Many youngsters and some unfortunate older animals hit on the road, run into fences, or otherwise injured or starving will be taken in, brought to health and maturity and taken back to the wild. The Highway Patrol and Sheriffs are particularly grateful for the Rescue Team resources.

In this case, Julia responded immediately and summed up the situation to be two very lost and hungry lion cubs who were wandering in town in search of food and company. They were obviously missing their mother, who should have been caring for them and teaching them how to survive in the wild.

Julia prepared to tranquilize the pair, bring in the care of a veterinarian, and nourish them until they were able to fend for themselves. Before she could act, Fish and Game bureaucracy asserted itself into the situation with the arrival of the local Warden, who stopped any tranquilizing and delayed until he could radio headquarters for further instructions. He was able to take charge as this was his jurisdiction. The Sheriff's deputies continued to provide backup and awaited events, and Solvang residents watched from their porches and windows.

At length the cats had enough of the impasse and leaped a fence to escape into the night. This was the worst possible development with two loose lions, even young ones, hanging around town. Julia, Warden, and deputies searched without luck and the watching residents wondered. By now the lions had become a media event, and the County began following the sightings and exaggerating the threats.

Late the next day, the lions were reported in a Solvang dumpster area on Fifth Street. The deputies arrived at the scene and immediately called Julia, who responded at once. Julia netted and tranquilized the young animals and brought them to the safety and veterinarian resources of the Animal Rescue Team shelter, this time before Fish and Game could interfere. But unfortunately the Warden appeared at the shelter and demanded custody of the unconscious pair. Fish and Game had the authority and would not answer where they would take the two, now caged and in an open Fish and Game pickup.

The lions were truly caught in a bureaucratic mess with the public, the media, and the shelter asking for explanations and Fish and Game simply refusing to divulge any plan for the two cubs. What is known of the initial handling of the two young animals was certainly less sympathetic than the Valley would have wanted. At first, one lion wound up in a northern Wisconsin zoo, and the other in a New York zoo, even though they should not have been separated. Fortunately, as of this writing, and hopefully for the rest of their lives, they are permanently housed together and prospering in a zoo in the Sacramento area.

Of course the Valley would not be happy to have lions on the golf courses or backyards, especially as they grew larger. However, everyone recognized that the good work of the Animal Rescue Team would have been better for the cubs than the heavy hand of government represented by Fish and Game and other authorities. They are secure in the zoo, but a life rehabilitated in the Los Padres Forest would certainly be a more desirable outcome.

Of course this story only accounts for two of the three lions originally seen by Jim Farnum. Small lion tracks have been observed on trails around the Alisal, so this lion story is certainly not over at the time of telling.

Another late-breaking turn in the story was announced in December 2009, and would certainly fit under the category of "No good animal deed goes unpunished," if the California Department of Fish and Game has its way. The Department filed a complaint against Julia Di Sieno for violating her animal rescue license, which did not include mountain lions. This could have resulted in both civil and criminal penalties. Fortunately our Santa Barbara District Attorney decided not to pursue charges, a decision very much in line with Valley sentiment in the matter.

· 𝒥 ·

Ugandan Visitor

ON A roundup day a number of years ago, a rancher and his cowboys were looking for stray cattle in a box canyon when an oversized, completely foreign, horned animal came bounding out of the brush to the total consternation of men and horses. Horses will put up with almost anything, but a sight like this stranger undid the animals, and one totally ran off with the cowboy.

When the dust had settled, the riders compared notes but had no explanation for the startling animal sighting. Days later the rancher was driving his Jeep in the same area of the ranch and saw the same strange sight, again with no explanation. What he saw was an African eland placidly lying under an oak tree, quite at home, although very much out of place.

Elands are the largest horned members of the antelope family. They have long, corkscrew horns, and enormous baleful eyes. They are gentle giants, and perhaps their placid nature accounts for the fact that they are the most efficient converters of grass to animal known to science, and they grow to over two thousand pounds. Elands are tan with a dark stripe down their backs and their mass and the lethal horns give them their serenity. Deer will run from anything and bound across the countryside, but elands tend to stand their ground.

Some discreet inquiries indicated that this eland was possibly the product of an exotic animal breeding farm in Los Alamos and had somehow escaped while in transit in the Valley and had decided to stop on this particular ranch. There were also suspicions about a zoo-type project that the creative mind of Vince Evans had envisioned for Buellton, but Vince made no claims. Fences are nothing to these beasts, and his cross-country journey must have been incredible. This eland had finally decided to give up traveling, and with abundant food and water and the cool shade of oaks had made peace with the world in quiet, albeit lonely, contentment.

The animal exuded good vibrations, and the rancher grew affection and respect for the beast as he observed the new boarder from a distance. The local ranchers had always had their suspicions about the exotic animal breeding operation. After a quiet conversation, all the cowboy witnesses to the visitor decided keep the matter to themselves, let nature take its course and let the animal stay put. They decided they could not bring themselves to stirring up animal parks, transit companies, and contented elands.

There is a natural inclination in the Valley to respect and work with animals. This eland was enjoying life and that was met with sympathy and understanding. The end result is that very few people ever knew about the eland, and it is only mentioned here because the animal and rancher are no longer with us, although confidentiality still needs to be respected. Africa was a long way away, but this eland was at home, and welcome for as long as it wanted to stay, which turned out to be the rest of his life.

· 𝒯 ·

No Hunting

THE SEDGWICK Reserve is a spectacular six thousand-acre ranch bordering the Los Padres National Forest owned and operated by the University of California, Santa Barbara. Up until the eighties, it was owned and operated by the Sedgwick family that included the famous Duke Sedgwick and his trendy daughter Edie Sedgwick. The descendants of the family helped the University acquire the property for research and retreats. The ranch is very much a preserve, and of course any form of hunting is strictly forbidden.

In addition to the usual authorities, including the Sheriff's

Department and Fish and Game, the UCSB Police Department also maintains security on the Ranch. A few years ago, campus Police Chief Dan Massey was patrolling the ranch from a high ridge, and spotted four heavily armed hunters working their way up the canyons toward his lookout point. He maintained radio contact with the Sheriff and Game Warden, but his dragnet only included sitting tight for an hour as he watched the poachers make their way to his concealed spot and confrontation with authority. Unfortunately for the hunters, they had bagged a small deer on the way, so the arrest and confiscation of all arms and ammunition needed little further evidence of wrongdoing.

More recently, two hunters won an even more inept poaching prize. In the plain light of day, a character from Los Angeles was spotted by a patrolman, dragging a deer carcass across the ranch. Obviously this city dweller did not understand Valley hunting laws and had ignored the signs posted to that effect. When he was stopped during his poaching foray he was happy to chat about his activities. He pointed out that he did not have far to go, only to his partner waiting in a van a little way off in the bushes which he showed the about-to-be arresting officers. The van yielded the partner, a bucket of ice cold beer, and a butcher knife waiting for the venison.

By sheer bad fortune for the poachers, the van was parked just off the road where a large sign advised visitors that there would be no hunting or firearms allowed. The city boys' sins included hunter trespassing, lack of hunting license, no deer tags, illegal method of take (using a .22 caliber rifle), and use of illegal lead ammunition in the condor protected area.

Unbelievably, the happy hunters agreed to pose for the officer's camera, with the deer victim in front of the No Hunting signs which resulted in a very expedited and conclusive conviction. Valley residents know that the Sedgwick Ranch is bad karma for hunting, and Valley deer know that the Sedgwick is very much the place to go for safety during the hunting season.

· *F* ·

Part Two
Country Moments

Morning Pals

A NYONE STOPPING for an early cup of coffee, newspaper, or gossip in Los Olivos at the "R" Country Store will encounter a friendly group of locals who sip coffee and solve the problems of the world. The gang that comes and goes at that hour includes some local dogs that are very much at home lounging around the store and hoping for Danish pastry parts. The cast of characters may change over the years but the comfortable ease of dogs and people remains.

Frank often brought along Wooley, who looked like and probably was mostly wolf, but had the benign disposition of a sloth and favored lying across the store entrance, the better to scare off tourists with his formidable if harmless presence. Gordon usually brought Chico, who was a medium-size friendly mixed breed, ever busy scrounging. Chico effectively used his dog eyes to promote a piece of Danish from any and all, and sometime the cold, wet dog-nose nudge to good effect. John sometimes brought Henry the Pug, and there have been, through the years, various other dog characters that add to the parade. The town cat, Mia, would slink around in the middle distance after accomplishing the nightly rodent patrol.

The easy atmosphere of dog and codger familiarity was broken one day in the spring of '08 when a lady, obviously having a bad day, was hurriedly ordering a sandwich from the counter. Unfortunately, Jamie was next in line, in the company of the friendly and unassuming Labrador, Bobo. "Dogs, dogs, dogs!" she exclaimed. "All I see around this place is dogs. Doesn't anyone know that having dogs in a food establishment is against the law? I am going to call the County authorities!" She paid for her sandwich and stormed out in a huff.

What to do? The store management, including owner Diane, who often brought French Bulldog Lilly and Australian Shepherd Maddy along, had no desire to meet the County health authorities and decided it was time to clean up the animal act in her store. The front door soon sported a large sign prohibiting all dogs, and legal conformity prevailed in our little town.

The law is California Health and Safety Code Section 114259.5 which states in part: "Except as specified, live animals may not be allowed in a food facility." The exceptions include edible fish, animals intended for consumption, dogs under the control of a uniformed law enforcement officer, service animals that are in the control of a disabled employee, and live or dead fish bait. Wooley, Chico, Bobo, and company were definitely out in the legal cold, and soon canine store visitations were banished to outside the store.

Fortunately, over time, Valley animal sensibilities prevailed. The proprietors were not arrested and it seemed the County had no interest in Los Olivos loose dogs. A few months later, if a stray tourist were to come by for coffee, they were once again liable to trip over Wooley and give up a breakfast roll portion to Chico just like old times. Somehow the complaining lady soon began to have much better dog days and ease with our Valley ways. And, by the way, if any California or County Health authorities are reading this, I hope they will note that the store in question is in Bakersfield or Visalia or somewhere out of state.

· 𝒥 ·

Mixed Blessing

BELIEVE IT or not, there is a formal, old-English-style hunt in our Valley. The members don formal hunting attire and follow a pack of fox hounds that scramble after a scent laid down across our rolling ranch hills to replicate a mad chase scene in the best of hunting tradition. The Huntsman controls the pack, assisted by designated Whippers-in, and the Hunt Master leads the Field to follow the hound action as closely and safely as possible. The riders of our Santa Ynez Valley Hunt are as well mounted as any local riders, ideally on big, strong Thoroughbreds that can run and jump and enjoy the sport for hours at a time. The hunt has been described as the ultimate trail ride, and is not for tentative riders. The only thing missing are the foxes, who tend to live in the pleasures of urban areas rather than risk coyotes and bobcats in the wild, and the bottled scent trail makes up for this lack.

After months of hound and horse training and preparation, in the late fall the Opening Hunt will be scheduled. This formal occasion calls for the best turnout of the Hunt members ready for a new season of galloping over the hills in good company and in good spirits. The riding sport is governed by the ancient formalities of hunts that have evolved over the generations and are faithfully carried out in our Valley.

One of the time-honored customs of the hunt is to bless the hounds and riders on the opening day, which sometimes occurs on a Sunday. It is not every cleric who will come out on the Lord's Day to bless energetic hounds and insane riders who are dressed in formal attire and mounted for the chase instead of singing hymns in the pews, but sometimes the church needs to follow a potential congregation, and the Blessing of the Hounds is an evangelical opportunity.

One year, and I can attest as a witness, a scene occurred that is straight out of an English comic farce, and is a standard of hunting cartoons. A local priest from the Santa Ynez Mission had originated in Ireland, where hunting and hunt blessings are a way of life. He agreed to come out on Sunday between services and do the honors for hounds and riders in an accent that evoked the best of Irish hunting tradition. He was a no-nonsense priest who had blessed hounds and riders in the Old Country and was well prepared to carry out the ecclesiastical formalities here.

The priest was trucked out to the opening meet where hounds, staff, and riders were assembled for the blessing ceremony. The good man in cassock, surplice, and stole mounted a rock and said a prayer for the safety of all and duly blessed

hounds and riders. As is also in the best Irish tradition, he stepped down, surrounded by hounds, to accept a stirrup cup and a toast from the Hunt Master. It was a picture-perfect scene, with milling hounds, staff and Field riders decked out in scarlet coats riding beautifully fit and well-turned-out horses, enjoying a sip of spirits before a great day of hunting. The friendly priest beamed on the hunt company, doubtless reminiscing on memories of Irish countryside.

Suddenly, a mindless hound lifted his leg and sprinkled the back of the guest of honor's cassock! Horror of horrors registered on the members of the hunt who had just received a divine intervention at the insult to the consecrated man of God. But the good man had not noticed, and the Master of the Hunt gave a glance to the Field members indicating that one snigger or recognition of the misstep was not to be tolerated.

With the usual style and excitement, the Master initiated the day with a call of, "Hounds Please!" The Huntsman blew his horn and the pack formed and paraded into a rolling pasture with Whippers-in at the sides of the pack, followed by Master and Field in high expectation. Soon the hounds were cast and began running a scent that moved the assembled company into a gallop across the Valley in a great scene of the chase. The priest returned to the Mission, unknowingly bearing an innocent mark of hunting tradition that, certainly the Good Lord, wise in the ways of old Ireland and crazy Californians, would forgive with Divine Amusement.

· *ℱ* ·

Chicken Dinner

THE SOLVANG Theaterfest is the pride of Valley entertainment, and great shows are held in the outdoor theater in the heart of Solvang. The theater was built by the generosity of Valley people inspired in the mid-seventies to construct an outdoor home for the Pacific Conservatory for Performing Arts, part of the Santa Maria Allen Hancock Community College.

Locals know that the theater is built over ancient Viking ice caverns connected to the North Pole that chill the seating during evening shows. Tourist visitors show up after a warm, sunny Valley afternoon in tank tops and shorts while the locals bring out their ski wear. The sun goes down, the temperature drops, and visitors flee to rent blankets while the neighborhood snuggles and enjoys the performance. In the early days of the theater, a number of us would stop by the Danish Inn restaurant after the show to warm up with late-night hot soup and other libations.

One day, a group of pals found themselves at the long banquette enjoying the aftershow glow and each other's company. The Danish Inn owners, Vince and Margy Evans, sat with Dean and Kathy Brown, Monty and Pat Roberts, and Tom and Maggie LePley. Kate and I were with John and Glory Bacon as well as other friends at the long table opposite the bar. Kate and I left as the party was still going strong, and the Bacons also said good night to the crowd. We walked out into the parking lot to discover a pickup-load of chickens in cages.

Wolcott and Nancy Schley had a small farm where a flock of chickens had grown numerous beyond their welcome. They were Rhode Island Reds and good layers, but the hay barn setting had caused a number of eggs to hatch and the Schley homestead was overrun with chickens. Tom LePley, a friend of the Schleys, coincidentally wanted to start a flock. To reduce his crowd Wolcott had snuck into the barn roost that night and plucked some sleeping chickens off their perch. He put the disgruntled hens in small cages that were deposited by prior arrangement onto Tom's truck parked at the Danish Inn to move to their new home. So far so good.

Glory Bacon was a sweet, lovely person who almost never thought of a wild, bizarre joke to pull on a pal, but somehow that evening she outdid herself. The four of us, Bacons and Firestones, were surveying the hens in the truck and something made her observe that Tom probably wanted to have a chicken join him in the restaurant. The Devil made me think that was a charming idea, and when Glory said, "Wouldn't Tom want to share with his chickens?" I replied, "Of course!"

I reached into a cage, pulled out a hen, tucked it under my jacket, and the four of us returned to the Inn for another nightcap. We sat down at the end of the table where Danish Inn owner Vince Evans was presiding. Vince was the bombardier of the famous Memphis Belle B-17 WWII bomber and had flown many missions in the war before becoming a screenwriter, restaurateur, and life of the Valley. He was usually ready for anything, but his composure went out the window when I opened my coat and freed the flustered Rhode Island Red to cut a flapping swath down the astonished table. Vince sat open-mouthed.

I have a picture in my mind today of Tom LePley registering shock, amazement, disbelief, and, fortunately, humor as I said, "Isn't that your animal?" People yelled, glasses flew, and shock and awe prevailed as the chicken careened down the long table and on across the bar and dining area. The staff went berserk, and for some reason of loyalty and defense of the Danish Inn, the cook came out with a carving knife to uphold the honor of the establishment. In short order, the hen was caught and disappeared while Kate and Glory soothed all the ruffled people feathers.

There is a haunting pause after one of those moments, when reaction hangs in the balance, and the perpetrator knows that the episode could tilt either way. This had the potential for a moment of audacious mirth or a damn fool thing. The chicken escapade had some things going in its favor. In those days the Valley had an easy country outlook. The Theaterfest had been great, the company warm and friendly, and the libations served as a receptive influence on reactions. To my everlasting relief, mirth prevailed and the assembly settled back in harmony of laughter. Vince leveled his gaze on me and said, "If you got any more of those damn chickens out there, bring 'em on in and we'll make soup!"

· 🍃 ·

Disaster Date

IN OTHER Valley Animal episodes, Charlotte Bredahl is described as our Olympic Champion dressage rider and Joel Baker as a seven-goal polo player and on the winning National Championship team. These are world-class horse people, but it did not seem so on their first date.

Joel was lucky enough to meet the gorgeous Charlotte and she agreed to go on a riding date with him. He knew of a remote location near the Guadalupe Dunes with access down a cliff for a romantic ride on the ocean beach. This was just the thing for two great riders, except that Joel, always practical, brought two green four-year-olds for a little training exercise, as long as they were taking the ride anyway.

They arrived at the top of the cool, windswept beach and the horses' eyes widened and their ears pricked up with a shivers of excitement. Charlotte took one look at her young horse and the cliff path and said thank you very much, but she would rather lead the horse down. So at least they got to the beach in good shape, but now the fun began, because these horses had never seen ocean.

Joel was still hoping to get a little experience and training into his young Thoroughbreds, so he suggested Charlotte take hers into the water. The horse snorted and balked, and she came up with the idea of backing him into the water so he did not see the approaching surf. Unfortunately, without warning, a wave came swooping in and inundated the horse up to his saddle girth. He ran deeper into the water and started bucking and soon had himself soaking wet and Charlotte in her beautiful boots and breeches out of the saddle and immersed in the surf.

But this was the first date, and Joel helped Charlotte back on as they both laughed at her saltwater drench. Then it was Joel's turn. This time, another big wave caught him riding in the water, and his youngster went over sideways, putting horse and rider completely under water. Finally the water-logged couple on their dripping horses set off down to the beach to the amazement of a few onlookers, who thought this couple probably had never been around horses before in their lives.

The wind and water began to feel cold, so the couple decided to return to the truck and trailer and call it a day. Once again, Charlotte took one look at the cliff and asked Joel to lead her horse up while she walked. It was a good call. Somehow the lead rope from the horse behind became caught under the tail of Joel's horse in front. Horses have no patience whatsoever for a rope tucked under their tail, and this one started bucking and promptly put Joel into the bushes at the side of the trail as both horses ran for the trailer. Joel's cold, wet, scratched, bruised, and humiliated condition began to wear on the joys of this first date. Charlotte shivered and pondered as they drove home.

However, fortunately for all concerned, any discomforts on the way home must have worn off quickly, as Charlotte and Joel have been happily married for many years. Beach rides have not been featured on any of their anniversaries.

· 𝓙 ·

Leash Law

A DOG'S LOVE for home is enhanced if the family owns a pickup truck with a back for the great pleasures of open-air riding.

A typical and happy Valley sight is a working truck with an earnest rancher at the wheel and a happy, wind-in-the-face dog perched in the back. The truck bed rider has a sense of mastery and adventure. The delicious wind rushes by with a smorgasbord of ripe aromas combining roadside kill, live game, barbecue, bitch in season, savory trash, and other delights to inspire the doggy mind. Meanwhile, Dad does all the work up front in the hot cab.

If there is absolute, defensible dog territory, it is the bed of the family pickup. The front belongs to people, but the back is dog turf. Sometimes there is a tool box or hay bale to enhance the perch and sometimes refuse for a comfortable bed. There may be two or three family dogs teamed up to enjoy the truck, but whether a single or pack, the back space belongs to the canines.

Anyone who speaks dog knows that only a good and well-known friend should approach a parked truck if there is a territorial occupant in back. Sometimes a well-meaning stranger will attempt to deliver an item to the truck and be repelled by the snarl and roar of the dog in charge. Even innocent roadside pedestrians are subjected to warning barks if there is even a dog suspicion of threat to the pickup. Often the dozy dogs wait until a stranger comes close to the truck and then let loose a startling bark, always with gratifying, panicked reactions.

There are a number of dog/truck protocols that are fairly predictable. For example, when a truck enters a neighbor's ranch, the resident dogs on the ground will bark and challenge, but will never attempt to jump into the back of a pickup, which is the visiting dog's territory. The dogs in the truck will return the verbal insults, but rarely if ever leave the truck to meet the challenging locals on the ground. Once the truck is parked, all will settle down and the quiet will remain until the truck restarts to leave. Then the ranch dogs will chase the truck out with howls and snarls, while the truck dogs return a cacophony of invective until the vehicle departs the property.

Once my friend and incredible ranch dog Daisy was riding on the back of a flat-bed truck with no sides for protection or containment. That was probably not a good idea, as it was drizzling and the truck was slippery wet. As the truck went around a corner, Daisy slid and slipped over the back and onto the road. A smelly, damp shepherd dog administered one of those piercing baleful looks as she was brought into the cab, and the driver was duly admonished. However, no story of a dog voluntarily leaving the back of a moving pickup, even for a meaningful chase, has ever been recorded, although it might possibly happen.

Most truck-riding dogs have memories of the way life used to be, before a well-intended but spoil-sport Assemblymember, who had the Valley in his district, blew his legislative whistle on dog/truck freedoms. This legislator, not the author, and who out of bipartisan accommodation shall remain nameless, was moved to help us manage truck-riding dogs' safety. This particular piece of legislation is typical of the Nanny government direction that is stealing our freedoms and independent judgment, but that is another story. A law was authored and duly passed to require dogs riding in open vehicles to be secured with a rope or leash to the vehicle. The dog freedom of the open road vanished!

Now it is conceivable that dogs like Daisy might fall out of trucks and become injured. And it is possible that some daffy dog might jump out of a moving vehicle causing injury to others or a traffic hazard. However, dogs probably do not think this is likely and surely resent the confining rope interfering with one of life's open-air moments. But the law is the law, and an untied and free dog in the back of a truck is cause for a citation if that is worth the trouble to a local Sheriff's deputy.

We have little idea of the true thoughts and dreams of our dog friends, even though we believe we can sense their minds. But one can imagine a vision in the dozy dog, sleeping in the back of his truck. Perhaps the dog heard the details of the leash law intended to confine moments of open-air joy. Perhaps the dog sensed the identity of that certain Assemblymember who inflicted the confining punishment on the dog world, and perhaps the dream might turn to thoughts of revenge. Perhaps, behind the half-closed dreamy eyes there might be a growling desire that the politician would walk by the truck, see that the dog was not attached and then, wanting to secure the despised leash, reach in, just close to the vengeful fangs. Drool!

·*ℐ*·

Elephant Emergency

THE TRAGIC death of Michael Jackson has removed some of the secrecy from behind the gates of Neverland Ranch, and information about the operation is beginning to emerge. These stories are new to the Valley because everyone working at the ranch has been under strict instructions and contracts not to reveal any details of their work. The fact that very little gossip from the ranch circulated over the years of Michael Jackson's life here is a tribute to the man himself, who inspired the loyalty of his employees. There doubtless will be many stories to emerge now, but the sources of this anecdote will still remain anonymous.

The ranch itself is one of the beautiful properties in the Valley. Personally, I knew it as the Easton Ranch and gathered cattle there in the early seventies. Bill Bone bought the property in the late '70s and built a substantial mansion on the property and carried out much work to improve the trees and roads on the renamed Sycamore Canyon Ranch. Michael Jackson bought the place in the eighties and it became the Neverland Ranch. With an imposing gate, security detail, and few local visitors, it became a mystery to the Valley.

The Valley did know that he improved the buildings and added an amusement park and miniature railroad, including an engine whose whistle could be heard over the hills. He expanded the barns and outbuildings to house a mini-zoo because of his own fascination with animals and to please his many visitors, young and old. He kept a stable with horses for his guests to ride, but never rode himself. Jackson usually wore a mask on the ranch to ward off any pollen or animal dust that could inhibit his singing voice. The zoo included giraffes, alligators, primates, reptiles, birds, and other exotics that were meticulously cared for, in addition to some that he rescued and brought back to health. People would give him animals, including a yak from Elizabeth Taylor and then an aged elephant also presented by her.

The elephant was a particular favorite of Michael Jackson's and he often visited and gave treats to the animal, which lived happily in an elaborate barn and special corral. Unfortunately the large creature was subject to bouts of elephant arthritis which can only be imagined in the scale that would affect a many-ton body. The ailing pachyderm suffered in silence, but the keepers and the special exotic animal veterinarian on call knew the problem and tried to relieve pain and stress. One night the elephant finally collapsed and the staff rushed to the scene to try and help.

When a large elephant goes down, the big animal will have trouble getting up, particularly if debilitated in some way. A downed elephant can develop complications and even die from circulation and respiratory problems, and the alarmed ranch people went on full alert. They contacted the special vet on long distance, a local vet on call, and Jackson's business manager, who was available twenty-four hours a day for emergencies. Her reply to the ranch staff was that this was Michael Jackson's favorite, particularly as a gift from Elizabeth Taylor, and that anything and everything should be done to save the animal.

Everyone understood that the animal must be brought upright, but that was not an easy proposition for the prostrate mass of pachyderm. The construction company that did regular work on the ranch was called to help, and front-end loaders and forklifts were considered and rejected because of the barn space and the elephant weight involved. Finally the construction company owner rousted an equipment company owner out of bed after midnight, and a heavy-duty crane was ordered and rushed from Ventura to the ranch as the only chance to save the elephant.

While awaiting the arrival of the crane, the staff and local vet fashioned a formidable sling they worked under and around the beast who was medicated to relieve the arthritis. The elephant was quiet and accepted the pushing and shoving, understanding that the crowd was trying to help. The construction team realized that the boom of the crane would not fit in the barn, much less the machine itself, so a hole was broken through the roof to lower the cable and lift the animal. At around two A.M. the crane arrived, the cable was lowered and attached to the sling, and the huge animal was propped up on its feet and stood, weaving as elephants do, but stable. By now the animal was full of painkillers and anti-inflammatory arthritis drugs, and able to breathe easily and chew hay standing up. The crew celebrated with coffee and rolls from the Jackson kitchen, for by now the entire ranch was involved in the drama.

The next day, under a shroud of secrecy, as was usual with Jackson ranch animal activities, the elephant was installed in the vet's clinic to the amazement of a few horses and clinic staff. After consultation with elephant specialists and a strict regime of medications and special feed, the big animal recovered and returned to the ranch. The hole in the barn remained, and the crane was on standby in Ventura, but fortunately was not needed for the elephant's remaining years.

In time, Michael Jackson lost interest in the ranch and moved overseas. There have been rumors about the treatment of the menagerie after the owner left the ranch and before it was sold, but the animals received only the best care and placement. The former staff, to the extent they are beginning to talk about the ranch years, are only complimentary about the detail with which each animal was found a home and cared for in the interim. The giraffes, alligators, and elephants have gone now, and Valley animal life is returned to normal and probably still wondering about the exotic animal noises and scents that emanated from the mini-zoo during the Neverland heyday.

· 𝒯 ·

Viva Outriders

ONCE A year, on a very special Friday, Valley riders wake up with the excitement and anticipation of riding in the annual Santa Barbara Fiesta parade. Some are destined to drive carriages, but most fun of all, and with less risk of disaster, is parading as outrider to one of the carriages. The event is one of the largest horse parades in the country, and a traditional day of fun for all of Santa Barbara, but mostly fun for the riders.

There are many horse-drawn carriages and floats in the parade and each entry will have at least two escort horses as company for the parade. The general idea of the outrider program is to fill out the street and add to the excitement and fun of seeing many horses, and occasionally the outriders also serve to calm and control the harnessed horses that pull the wagons. Occasionally, outriders prevent carriage runaways, as unfortunately happened in 2009, and which could be a real disaster in the crowds. Outriders are supposed to have calm horses, capable of stopping and standing still, that do not mind a hundred thousand kids sitting on the curb and throwing flowers, and are happy with marching bands and dancing girls with a few flags thrown in.

In the 2008 parade, five riders were chosen to accompany the Grand Marshal of the parade, our own Valley's Cheryl Ladd. Cheryl is one of the most pleasant and natural as well as most gorgeous neighbors who stars in film productions and TV shows like "Charlie's Angels" when she is not doing good works in our County. She was perfect for the Fiesta Parade Grand Marshal, and there was no problem whatsoever in rounding up outriders. Tom Le Pley, Steve Golis, John Cochrane, Larry Saarloos, and I were nominated as Grand Marshal carriage escort outriders and showed up on time, well cleaned and polished. Cheryl's very pleasant and understanding husband, Brian Russell, showed up also. Since I am one of thousands of locals in love with Cheryl, I greeted him with a "Sorry you came along!" He replied with the standard, "That's why I am here!"

Parade Chair Wayne Powers and his committee supervised the nervous moments when riders mounted their parade horses and tried to line up in the confusion. Every parade has its disasters, with nervous horses running off, tack breaking, riders falling off, or whatever, but this year, everything seemed under control. We formed up with Tom and me at the head of the carriage driven by John Tracy, and Steve, John, and Larry riding behind. Cheryl and Brian sat in the carriage, and our unit moved out, near the head of the parade. We had the Sheriff's unit ahead and a carriage-load of City Councilmembers behind.

The first few blocks are easy because Cabrillo is wide and the crowd is light. There is always a start-stop situation, and John Tracy asked me to put my horse directly in front of his two mules when we stopped to keep them from fretting. I was well mounted, but Tom had a more nervous Thoroughbred with an obvious missing eye. Someone in the crowd asked him about how a one-eyed horse went, and Tom replied, "Okay, because he is blind in the other eye and it evens out!"

When we turned onto State Street the fun began. First there are the tracks to cross that look like cattle guards, and some horses always refuse to walk over the railroad easily. In 2009 a horse reared here and put its rider in the hospital. Then there is the down-and-up below the 101 overpass, and someone always slips there. Then there is the daunting view of State Street with thousands of people waiting to see someone fall off. Cheryl, of course, got all the attention in our group, but that did not keep the kids from throwing flowers and the occasional balloon at our horses. A horse shying on the crowded street has little traction and nowhere to escape to, but we soldiered on. We always seemed to stop where there is a loud speaker blaring into the horse's ear, which is a signal for prancing.

On some of the stops Tom and I tried to inspire the crowd with a *"Viva la Fiesta,"* and the crowd responded with the same. Then the boys in the back did the same and tried to outdo our crowd. Pretty soon the press of people became heavy and I always wondered how small kids sitting on the curb with their bare legs a few feet from the horse's hoofs in that nervous atmosphere are able to avoid disaster. Additionally, there are many nearly topless beauties to steal the concentration from outriders. But somehow, as in every year, we came to the end of the parade unscathed, heaved a sigh of relief, and rode the long trek back to the park.

The horseman's party is held after the parade at the Carriage Museum. It is one of the best parties of Fiesta, because only parade participants are allowed, and everyone is happy and relieved to have survived another year. We outriders sat with Cheryl and in the course of conversation the truth about her escort riders came out. Chairman Wayne Powers had made a big point of coming over to congratulate our gang on being well-behaved outriders. The fact was that two of our number the previous year had held up the parade while they sidestepped around the street, flirted with the beauteous bystanders, and then, an absolute sin, roped an onlooker from the crowd. They were very much on probation this year, under threat of never being allowed to ride again. Cheryl squealed and was absolutely delighted to have outlaws for outriders.

· 𝒥 ·

Night Hunting

NIGHT HUNTERS are a unique subculture in our Valley that enjoy crashing around in the dark countryside, following their beloved hounds wherever they may lead until the pack has put something up a tree as a sign of surrender. Sometimes a wild boar might come to bay, and that can become exciting when the boar becomes the hunter and people climb the tree. Night hunters love the chase, and rarely kill anything because they never find anything worth eating and they want to chase whatever they find again. At one time there were a few night hunters in the Valley, and may still be, but they are a shy group, and not featured on the covers of sporting journals.

The hounds are best pictured as lanky, floppy-eared creatures who live under rural porches. Only those very much in the know understand how talented these creatures are, and how a keen nose and deep voice is prized. When not on the chase, they tend to be very affectionate and only mildly territorial and also a little shy.

Midland School is located on a large, beautiful ranch that is integral to the school's history and culture and has never allowed hunting of any kind. The ranch is ideal for night hunting, though, and one of the Midland faculty in a moment of lapsed judgment was talked into going out night hunting on the sly by a hound enthusiast friend. (Names withheld by request.) On the designated evening, a few night hunters and their loyal, hardscrabble hounds met with faculty members on the back of the Midland property in the middle of a dark night, long after Headmaster Ben Rich had turned in and with the campus quiet and peaceful.

The hounds were set off into the unknown darkness, and the hunters, bundled in their warm clothes and fortified further by spirit-sustaining libations, set off in stumbling pursuit. The distant moon cast dark shadows, and the night quiet was occasionally interrupted by the rustle of hounds hunting in the bushes.

The adventure proceeded with the hunters following the direction they believed the hounds had taken, hoping for the sweet music of a hound crying out to his pals that a scent has been found and telling them to come on over and join the fun. Hounds are pack animals and like to hunt in company, so will rally to a speaking voice in the night. When they are in full cry there is a true dog Hallelujah Chorus. The hunters will then move as fast as possible to follow the pack and be in on whatever transpires. A bobcat up a tree with cat's eyes staring down into a flashlight is a totally rewarding evening. Raccoons or possums are fun, but a wildcat, with the rare possibility of a mountain lion, is a night well spent.

On the evening in question, the hounds seemed to be working in the general direction of the quiet campus. Finally they all opened in full voice with the sweet music that only a pack of hounds on a quiet night can produce, and the hunters harkened and thrilled to the chase. Hounds continued to run the scent, then lose, and then circle and regain the quarry, generally in the direction of the campus. Soon they ran straight out, closing rapidly on the outbuildings. And then they successfully treed their game, but in the middle of the campus!

The hunters shined a light up the tree and, horror of horrors, they had treed Headmaster Ben Rich's tomcat!

Confusion reigned. Lights blinked on in campus windows. Night hunters leashed their hounds and disappeared and the guilty hunting faculty members melted into the shadows and vanished. The tomcat snarled and pouted from the branches of the tree. Headmaster Ben Rich emerged in robe and slippers shining his flashlight tentatively into the night. An innocent silence prevailed, which continued through school assembly next day and thereafter, as the episode of the strange disturbance in the night became a part of Midland lore.

· 𝒯 ·

Flying Horse

STEVEN REEVES and his wife were riding on a steep and narrow trail in the Los Padres Forest when the horses shied, backed up, and suddenly his large gelding fell about one hundred fifty feet into a deep gully. Fortunately Steve was thrown off onto the trail and did not take the tumble with the horse, and his wife and her horse stayed upright on the trail as well. They were two miles from anywhere, but with the help of cell phones, the Forest Service and Search and Rescue folks soon located them and the stricken and stuck horse in the bottom of the ravine. Then the problems began.

The horse was at the bottom of a steep, brushy ravine, well entrenched in a washed-out cut with no trail or purchase to gain traction and move out. The rescuers tried to dig out a path or at least some room to free the horse, but were completely frustrated by the terrain. The stuck animal was sedated and fed and watered for two days, and seemed docile and resigned to immobility. Unfortunately, this could not last, and soon everyone came to realize that the animal would need a miracle or, sadly, be put down.

As a last hope, the authorities contacted the Sheriff's Department Rescue Helicopter Service, and pilot Dave Wight and crew chief John Simon got the call. Sheriff's helicopters are used to put out fires, rescue lost hikers, chase down bad actors, discover illegal crops, and just about any other task that can be accomplished from the air. They had never rescued a horse. The two went out to the site and met the pathetic animal and the distressed owners. Could they do it? This service, of course, firmly believes they can do anything.

On the day of rescue, they flew where no sane person wants to go. The helicopter gently descended into the box canyon above the ravine, below hills on either side. A hundred-foot super-strong cable was let down to the waiting crew below. Meanwhile, the horse had been well sedated, blindfolded and trussed firmly into a sling structure that would support the animal in flight. Legs were wrapped and various cushioning devices shielded the horse from bumps and strains.

When the moment came, the cable was attached and the horse gradually ascended from the ravine, limply draped in the sling cage. The assembled rescue people cheered. The sedated and blindfolded animal slowly flew the two miles to the Upper Oso Campground and into the waiting arms of veterinarians, rescue workers, media, and the Reeves family. The pilots, Dave and John, deservedly felt the glow of a rescue job well accomplished. The type of people who accomplish these things tend to shrug off an incident as this as just what they do, but to the rest of us, a feat like this is a miraculous demonstration of nerves and skill.

By now the horse had become a significant cause and celebrity in the County, but with a gentle let-down and four-hoof landing, he just stood and awaited events. Events were unhooking, unwrapping, blindfold removal, kind words, feed and water, and a happy return home with a lifetime of celebrity as one of the few truly flying horses.

· *J* ·

Lethal Crossings

ANYONE WHO drives Valley roads knows that animals suffer road accidents at a horrendous rate. It is hard to run a Valley errand and not see the evidence of deadly car/critter contact lying on the byways.

One can only imagine that young ground squirrels are taught at an early age that the best fun is to wait at the side of the road for the thrill of dashing in front of speeding vehicles. But they are not taught very well, and most of us have felt the thump of squirrel tragedy on our wheels. The birds that make their meal on those unfortunate squirrels face a similar risk, desperately flying at the last moment from approaching cars and too often their indecision results in a flurry of feathers in the wind.

The Highway Patrol is tasked with removing a larger carcass if it presents a road hazard. Sometimes late at night, when the officers are riding in pairs, they might stop to remove a dead dog in the middle of the road. They carry a small rope with a noose to catch a leg to accomplish this. Night shift humor will initiate a junior officer when approaching a dead dog on the road by the light of the patrol car headlights. A senior officer behind the wheel has been known to switch on the car loudspeaker and let out a loud growl as his rookie partner is about to fix the rope.

Road clearance is also a problem for Caltrans crews who have road clearance responsibilities and are constantly called to remove a large deer or cow and sometimes a wild boar from the right of way. Policy calls for the crews to remove the animal to a County dump or rendering plant, but this takes time and effort and requires vehicle cleanup that is distasteful and interferes with more productive work. However, on the Valley byways the rules will sometimes be overlooked, and the carcass will be rolled to the side of the road and into a ditch. The many Valley animals that make their living off cadavers very much appreciate this breech of policy and the Caltrans people know that they are contributing to the food chain and Valley animal prosperity by ignoring the guidelines.

Every rancher lives in fear of a cow breeching the fence and wandering onto the road to cause an accident. Errant cattle meandering the road have potential for considerable liabilities to the owner if they cause a wreck, which sometime happens. For twenty-five years, Kate and I lived on a ranch that fronted Highway 101 for two miles. We kept an extra-strong fence and a constant watch and never had a wreck, but were always worried. Somehow the authorities knew I was available in the middle of the night, and I often got the call to help with loose cattle on that stretch of road.

At about two o'clock one morning I received a call from

the Highway Patrol and dressed and drove out to help put a loose cow back through the fence. I knew how and where to open my fence to readmit a wandering animal on the road and the patrol appreciated my help in eliminating a potential hazard that might lead to an accident. On this night the officer and I found a panicked heifer that had already been hit by a car and broken a leg. Cattle, if they are really spooked, can move fast on three legs, and this one dashed back and forth across the dark highway. A car just missed the running animal, and we could see other cars hurtling by with the potential for a high-speed wreck. We tried to corral her and coax her back through a gap in the fence. This went on for an exhausting half an hour, with some narrow misses from speeding cars in the darkness.

Finally the Patrolman asked if we could really accomplish anything. I replied that I did not think so and that the damaged leg, much worse by now, would probably doom her in any case. When the moment was right and the heifer paused for breath, he reached out of the patrol car with his service pistol and placed one lethal bullet in the poor panicked head that dropped her dead. I remarked to the Patrolman that I was glad we were on the same side. Before he departed on his rounds, I made the officer inspect the cow, now down by the side of the highway, with his light to determine that the "F Bar" brand that denoted my herd was not on the heifer's right side. If there were any liabilities from the leg accident they would not be mine.

Years earlier, just north of Buellton, there had been a serious accident caused by a cow on the road involving a number of cars. Names withheld on this one. Even in the late hours, a crowd collected, and some ranch hands helped pull the obstructing cow carcass off Highway 101 as wreckers removed the cars and ambulances removed the injured. An investigation was held, but the responsibility could not be determined. The next day, a ranch hand quietly presented the grateful owner with a patch of hide, containing the ranch brand. He had neatly removed ownership evidence from the animal with the quick stroke of a pocket knife.

Jake Willemsen moved to the Valley in the fifties and

established a dairy herd west of Buellton on Highway 246. In his big years, he worked up to two hundred and fifty Guernsey cows, but it was always a hard business. Finally, in 1989, he discontinued the herd and there are now homes on most of his pastures. One dark and foggy night in the seventies he had a call from the Highway Patrol that he had cows on the highway. Jake searched, and at first could only hear the clopping of many feet in the fog, but finally he got behind a couple of dozen cows that were searching for greener pastures in the night and brought them home. Milk cows will respond and move more easily than beef cattle because of their pleasant nature and as they usually expect feed or udder milking relief when they are brought home.

Jake looked forward to his warm bed, but the Highway Patrol Office told him he had another one down just off the highway. Jake inspected and found that one of his huge Guernsey bulls had been badly hit by a car and would probably not survive. He sadly put the badly wounded animal down and fetched his tractor to drag the large carcass back to the farm, now destined for the rendering plant in San Luis Obispo. He wondered about what had hit the bull and hoped that there would be no repercussions from the accident that could be financially disastrous.

The next day, by pure chance, Jake's wife, Jeanette Willemsen, was in Nielson's grocery store, then located in Buellton. In the checkout line she overheard a lady talking to a friend about the horrible accident she had experienced the previous night while driving in the fog. The shopper thought she had hit some big animal because of the considerable damage to her Volkswagen, and she only hoped that she was not responsible for any liability to the animal owner. Janet, pretending not to notice the conversation, concentrated on the middle distance and never said a word.

Valley animals do not understand our need for rapid transportation, and Valley drivers need to understand animal pace and road innocence.

· *J* ·

Nanny Horse

VALLEY HORSE people know they can find a home for their surplus, retired, or otherwise extra animals at the Midland School. Any Valley horse that is not too unruly or debilitated is welcome to join the school string and, horses being adaptable creatures, will usually fit in. The Midland School is blessed to be on a two-thousand-acre ranch, and a large field in the daytime and smaller paddock at night provide home for the horses all together in a herd. After a new horse is tested by an experienced rider, a student will adopt the horse for a year and be responsible for the care of the animal. Midland grads carry a lifetime of memories of happy days riding the Midland trails, and the students benefit from the responsibility and learning experience of a campus horse friend.

Midland prides itself on a cold-water, country atmosphere, where students build their own desks and live a very Spartan ranch school year. This atmosphere is not for every city kid, but those who buy the program have a first-class experience and education, graduate with great character, and mostly go on to be extremely happy and productive people. Horses play a good part of that scenario.

The horse population lives off the grass in the large pasture and hay thrown out in the paddock at feeding time. Additionally, the most favored food is the daily dining hall garbage laid out for the herd in the corral once a day. Horses will eagerly await leftover salad bar, cantaloupe parts, oatmeal, and other scrap items, and there is much milling about and pushing with the occasional horse squeal to be first for the favored morsel when the kitchen cans are dumped.

We do not know what goes on in a horse's mind. Some say that horses only learn habits, go on instinct, and are too dumb to think. Some say there is a lot more going on between the ears than we realize. Horses can be mean or friendly, competitive or lazy, willing or stubborn, and all other variations of horse traits, but why and for what wooly equine reason remains a mystery. The greatest of horse mysteries is what makes a runner want to win a race on a given day. This is the subject of endless speculation, and whoever figures that one out will be a rich person. Cowboys will sit by the hour psychoanalyzing steeds they have known, but end up admitting they just do not understand their horse friends. And then along comes an event to further confuse the subject.

A number of years ago, Jim Munger, then the teenage son of the Midland Headmaster and later a distinguished Dunn School Headmaster himself, was walking across the Midland ranch campus. He noticed that the garbage had just been put out in the corral, and he further noted that a big palomino gelding named Sig was standing unusually still in the center of the milling horses.

Sig was a large mature horse that had weathered a number of student years and was a popular fixture on campus. He stood well over sixteen hands and sported a bright blond palomino coat. Sig knew the joys of the garbage can well, and it was peculiar that he would be standing still at feeding time and not shoving in with the rest for a favored tidbit. Jim was wise to horse vibrations, and this did not seem right, so he entered the corral and waded in to Sig through the other horses and made an astonishing discovery.

Sitting quietly between the large palomino legs was Jim's three-year-old brother Ben, gazing out at the other milling horses from the safety of Sig's legs! The big horse was standing stock still, protecting the young child who quite happily accepted the safe haven. How he got himself in the middle of the milling herd was a mystery, but he was there and his situation had a very dangerous potential. Jim spoke to the horse and stroked him while taking up the child into his arms. Sig then, with a blank expression, happily joined the feeding in his indifferent and normal horse way. Jim left the paddock wondering.

This bizarre incident is just the way Jim Munger related the story in amazement many years later, and it has stuck with him as one of the bizarre moments of his lifetime with animals. An incident like this only adds to the mysteries of horse psychology.

· 𝒥 ·

Bike Encounters

THE VALLEY is good country for bicycles, with hard-core training teams, weekend long-distance riders, locals tooling along for their errands, and kids cruising the neighborhoods. Locals are accustomed to seeing many bikers on the sunny rolling roads, and these quiet, speeding cyclists inevitably encounter Valley animals to the mutual surprise of both.

Once I was driving down Zaca Station Road and three bicycling couples, closely clustered around an exceptionally large and hairy tarantula, flagged me down. I was in my pickup and looked like a local, so they asked me about the spider. I told them that the giant tarantulas gave a painful but usually not fatal bite and could jump eight feet. I found out that bicyclists could jump also.

Corey Evans owns Dr. J's bike shop in Solvang, leads bike tours, and has seen many bike and animal encounters but has never heard of a resulting injury to the cyclist. Rattlesnakes on the road, deer dashing across, and even the occasional wild boar have supplied biking tourists with stories for their urban friends. And of course Valley dogs are a constant biker's threat, but a defensive squirt from a water bottle will usually discourage the barking dog chase.

Ground squirrels enjoy waiting at the side of the road, until a biker comes by to see how close to the wheels a panicked rodent can come. Corey knows of one squirrel that was hit by a bike's front wheels and somehow was sucked up by the back wheel and dismembered on the spokes and brake mechanism. The biker survived, but the cleanup was long and noxious.

Sometimes cattle will push through a fence to graze on the lush grass to be found on a roadside, and often the agile calves will negotiate the fence and leave their mothers behind in the bare pasture. Corey, riding with some athletic friends, came on such a calf on Armour Ranch Road, about a quarter-mile from Highway 154. The calf took off running toward the busy highway in front of the bikers. The group thought they could sprint around the calf and then stop and keep the youngster from wandering into traffic, but the calf took off at a dead run and they raced to the highway.

The calf proved faster than the bikers and to everyone's amazement shot across the crowded highway where the Ranch road becomes Highway 246. Once the bikers negotiated the highway crossing, they set out in earnest and were able to get around the loose calf and head him down Meadowlark Road. Fortunately, a cowboy driving by opened a gate and the animal was penned until the owner could be found and the calf reunited with the mother cow. The highway drivers were treated to a little bit of Wild West action but luckily avoided a calf wreck.

Once, some of us sitting in the Bulldog were amazed to see three small dogs on separate leashes parading down the sidewalk in Solvang hauling their master on a bike. That and the occasional dog or cat in a bike basket is part of the unsteady relationship between Valley animals and bicyclists.

· 🦴 ·

Part Three
Commercial Enterprise

Double Larceny

A GENERATION AGO there was a pig farmer in the Valley who had a reputation for getting everything he possibly could out of his farm. There was also a bright teenager who was ambitious to buy young pigs, feed them for a weight gain, and sell the animals at a good profit. Both traders will remain anonymous, but this is a true story.

The teenager borrowed his father's ranch truck to pick up a load of the farmer's pigs. The two had previously concluded the price per pound after a good amount of haggling. The other part of the equation would be the weight of the young pigs to determine the money owed. The price negotiations were hard and fair, but the chicanery started with the weight factor on taking delivery of the animals. Pig poundage to calculate the total price would be determined by the weight of the loaded truck, containing the pigs, minus the weight of the unloaded truck.

The farmer, wise to the ways of pigs, had given them a very prosperous morning. Knowing the time the young buyer would pick up the animals, he had calculated just the right moment to give them the feed of their lives. The young pigs, accustomed to the farmer's parsimonious ways of feeding, had been surprised and delighted by unlimited quantities of sweet mash and had gorged themselves. Afterward, and shortly before teenager and truck arrived, the now-thirsty pigs were treated to a fresh, flowing watering trough. The pigs were happy, bloated, and heavier than they'd ever been in their lives, with the feed and water now being very much part of the swine sales weight.

The buyer was very junior to the farmer, but wise beyond his years in the ways of farming, and noted that the animals to be purchased were swollen and dozy, possibly in a well-fed and watered condition. He took on the load with the agreement that he would weigh in and out at the official public scale and settle up that afternoon with the certified weight slip. He took off down the road, but pulled over to the side and rested in the shade for a half-hour. At the end of that pause he had the satisfaction of sweeping out porcine pee and poop that he would never pay for. He called the farmer and reported he was delayed by a flat tire but would return shortly.

Next he carefully let the air out of a tire and limped his truck with the near flat into Steve's Wheel and Tire in Buellton. Here he left the low tire and had the spare put on. He also left the heavy tool box and other loose items from the truck at Steve's and drove off to the public scale. After registering the weight of the truck including the load of pigs with the scale operator, made lighter without the spare or extras, he drove to his family farm and offloaded the newly purchased animals.

On the way back to the scale operator to weigh the empty truck, he stopped by Steve's and picked up the tire, which was now a spare, without Steve having found the flat problem. The young farmer loaded the tool box and other items and also topped up the gas tank before weighing in. At the scale he weighed the now-heavier truck and received the official weight slip, evidence of the pig portion, which was the difference between the truck loaded with pigs and unloaded. The difference between the first scale pass and the second pass was the weight of the pigs to be purchased, but of course he had weighed first with a light truck and later with a heavier truck.

Our young entrepreneur presented the official weight document to the old pig farmer, who was somewhat disappointed in the weight of his pigs, but content in that he had gotten the best of the boy by selling water and mash for pig weight. Of course our teen-age buyer knew better. Two larcenous generations of farmers were pleased with the deal for very different and very secret reasons.

· *J* ·

Model Horses

ALAMO PINTADO Road runs down the middle of the Valley, and just south of Ballard, locals and tourists have enjoyed watching a miniature horse farm emerge in recent years. As a testimony to the popularity of this animal addition, there is a sign on the fence stating, "Please do not feed fingers to the horses!" The splendid little creatures are a delight to everyone driving by.

Around twenty-five years ago Aleck and Louise Stribling attended a miniature horse show at the Earl Warren Show Grounds. They were horse people and were quickly drawn to the engaging miniatures and moved to purchase three mares for family fun and companionship. Two foals from those mares, Taffee and Cookie, happily moved into the Striblings' household and worked their model horse magic. The herd soon outgrew Montecito and today their twenty-acre Quicksilver Ranch breeds twenty-five to thirty foals a year and the little horses have pretty much taken over the lives and property of the Striblings.

Miniature horses, smaller than a Great Dane but larger than a breadbox, have an ancient lineage originating in the royal houses of Europe. Their small horse charm has graced some of the great estates, and they now serve as companions and hobbies for many around the country. Their personality and behavior is similar to honest and straightforward horses rather than ponies, who sometimes have quirky ways. Miniatures are almost all registered and can serve as cart animals or show horses or simply great horse companionship without the scale and hassle of a full-grown animal. And of course they can be bred, and the herd expanded as the Striblings very much found out.

Many Valley visitors stop by the familiar Quicksilver Ranch on Alamo Pintado and some of them take home foals in the back seats of their vehicles. One lady stopped by with her daughter on her way to move into a new home in Palm Desert and bought a foal then and there to take with them. When the Striblings learned that they were driving a Ryder rental they explained that they could not release a foal to ride in the back of an enclosed truck. "Oh no," the lady exclaimed, "She will ride up front with us!" The little horse proudly rode to a new home in air-conditioned splendor between mother and daughter.

Miniatures seem to have an instinct for befriending young children and people with some physical challenge. Once a young girl in a wheelchair, severely affected by cerebral palsy, was brought in with a group of visitors to the ranch. A young miniature made his way to the girl and put his head in her lap. She registered the most joy they had seen on her face in months, to the delight of the caregivers. Often the small horses are groomed and certified as therapy animals to visit hospitals and care homes.

Years ago, Connie and Marsh Ream moved to the Valley

to establish the Zaca Mesa Winery. Marsh was retired from Executive Vice President of Arco Oil, and Connie, a native of Virginia, had been struck by polio in her mid-twenties and had lost the use of her legs. Prior to that, she was an accomplished show rider, and she missed her horse friends from the confinement of a wheelchair, although they lived on a large ranch with their five children, who enjoyed riding. A blessed event came into her life when she imported the first miniature horse. Soon she had four horses living in a specially constructed paddock out the door of her bedroom. The horses had a small miniature barn and every luxury, including the almost free run of Connie's bedroom.

Her horse involvement, which would never have been safe with full-grown horses, was now fulfilled with family miniatures, whom she loved from her sitting position. One day she overdid the closeness to her animal friends when, in helping birth a miniature mare, she fell out of her chair while assisting the process. In current days, Connie is living in South Carolina and maintaining her interest and memories by writing newspaper columns, including stories of her Valley ranching and grape-growing days, which very much include her miniature horse friends.

Valley types who drive by the Quicksilver Ranch might do themselves a favor by taking a moment to stop and visit the miniature-horse–enchanted world. But all should be warned that such a visit might lead to a Stribling type enthusiasm for the winsome animals—and ultimately a ranch takeover.

· *F* ·

Animal Inspirations

MANY A great career has started off with farm work. The realities of husbandry inspire young people with potential to impressive efforts in order to enter almost any profession other than the barnyard. Valley animals have helped motivate our community young over the generations and continue to do so.

Richard Christensen has had a long and distinguished career running the foremost accounting firm in the Valley. He has dealt with important local and international clients and estates and guided many families into sensible economics as well as keeping the IRS happy with their statements. He is retired these days to boards and trusteeships from his successful firm, and the Valley is grateful for his wise, professional, and able talents. But he wasn't always that kind of person.

In the 1950s, a sixteen-year-old Dick Christensen, full of energy and teenage ways, worked for the Burchardi Dairy. In those days, the pastures and dairy were in the vicinity of Highway 246, just west of Solvang, where many beautiful homes now occupy the space. Dairy work is tough, because the cows must be fed, milked, and cared for, and there is no putting off or slacking from the very real needs of the herd.

Dick was on duty by himself one hot day when a cow went into labor and was having major trouble in calving. She had taken station in a pool of cow muck in a low barnyard spot, standing firm with a calf halfway out of her, but obviously stuck. As he was working by himself, there was nothing else to do but move the cow to higher ground, out of the swill of cow pee and manure, and help her with the birthing. But no amount of coaxing or threats would move her a step forward or back. Finally the young cowherd waded into the muck to do his duty.

After a few minutes of pulling on the emerging legs, the calf was more than halfway out of the cow, and the birth finally began to happen. But when the calf was mostly out, the slippery newborn with Dick's pull on the legs reached the critical exit point and gave way suddenly. Without anything to hold onto, Dick found himself upside-down, totally immersed in the muck, with the lively calf and afterbirth on top of him. When he sat up in the dank, liquid mess, the cow added insult to the moment by turning and bellowing a loud cow moo, possibly by way of thanks.

Fortunately his other chores were over, and Dick was able to leave cow and calf to their happy domesticity and hose himself down. His clothes still couldn't be worn, and he did not want to stink up his car, so he set off on foot for the mile or so back to his Solvang home, wearing his shorts. He had bundled his smelly clothes and tied them to a stick, which he carried over his shoulder. Of course there was little traffic in those days, but the walk was long and humiliating.

During his tramp home, he had a hard and decisive think about his future, and the advisability of farming. Once again a Valley animal had provided the inspiration for a great future. With a determination engendered by the real farm world, Dick went on to college and grad school, and armed with CPA credentials began his long and distinguished career, but not in farming.

· *J* ·

Bulldog Brew

IN 2002, Barbara and Greg Meeks were sitting at home with their two sons, engaged in a serious discussion regarding their new business enterprise, a Solvang coffee shop. Most details of the project had been ironed out, but they had yet to decide on the crucial question of a name for the new venture. Comfortable on the rug, at home with the family, and probably dreaming of buried bones, their family Bulldog named Tank peacefully dozed. Perhaps he snorted or rolled over and somehow inserted himself into the decision, but for whatever reason, one of the boys said, "Let's name it The Bulldog!" and a Valley animal had again woven himself into the fabric of our community.

Tank joined the Meeks family from a Bulldog breeder, who had produced a stellar litter except for one, who was the classic runt. Somehow gestation had treated this tiny member unkindly and he came out something less than half-size. But there was something about the lesser one's personality, even in his early puppy days, that caught Barbara and Greg's eye. They had never wanted a show dog and felt a surge of emotional attachment to the little creature, smothered by his littermates and in need of a home. The mighty mite joined the Meeks family. He acquired the name Tank, settled in as part of the household, and received the kind of love only a Bulldog can inspire in the eyes of the breed's admirers.

Anyone who desires a good hot cup of coffee, healthy fresh-baked stuff, newspapers, and gossip makes a path to The Bulldog these days. Early on, a quiet and determined bicycle racer named Lance Armstrong found The Bulldog while training in the Valley and has been a customer ever since. His presence set a pattern for many bikers to park their precious machines outside the shop while stoking up for the ride. The world goes by on Highway 246, but inside there is a warmth of hospitality in the spirit of a plucky and comfortable Bulldog.

A large portrait of Tank slathering over a cup of cappuccino by Valley artist Tyler Gildred looms on the wall over the counter. A Bulldog statue outside serves as a dog leash fixture, and there is always a dog water dish waiting. Local dogs know that there is a jar of dog treats available on the counter and try to influence their people to linger at the coffee shop and then share. Local people know they can find out the latest Solvang story with their brew, and sometimes show up just to insert their version of the gossip into the mix.

Tank passed away in 2008, but lives on in the spirit and name of the shop. Once again, an animal has gifted our Valley with a presence, in this case with a benign dog coffee shop atmosphere that enriches all our lives.

· 🦴 ·

Wookie Drama

A COMBINATION OF luck, disposition, emotions, and perhaps a little thought determines who in the Valley will respond to and become involved with certain animals. Some gravitate to Arabian horses, some to dogs, some to ostriches, some to 4-H projects, and so on in great variety, each to their own.

Anyone who stumbles on alpacas to be animal friends has made a lucky decision. In the late '90s, our daughter Hayley acquired two bred alpacas as a project. In ten years the herd grew, with outside boarders, to include 120 alpacas, living on a piece of land owned by our winery. She and her family fell in love with the animals and obviously the relationship has thrived.

Alpacas are a South American animal similar to llamas, but smaller, kinder, and gentler. Their dispositions are universally benign and they grow a fiber that can be made into warm and wonderful clothing that has kept the chill off Peruvian mountain people for centuries and has also supplied the fashion houses of the world. They eat little, poop in one place, and have soft feet that leave no pasture impressions. With teeth only on the lower jaw and no inclination to bite, they are perfectly safe to be around, although occasionally they may spit. The wooly creatures emit comfortable noises and stare with enormous gentle eyes. Alpacas are so universally pleasant that they seem to come from another planet, and that is why our family calls them Wookies, out of the *Star Wars* strata.

Alpacas pretty much take care of themselves, and Hayley administers to the herd easily with the help of a cleanup person and an occasional visit from the vet. When birthing time comes along, they just do it, producing adorable littles called *crias* that are soon up on wobbly legs and then dashing around the pen in days. The big work comes with the annual shearing, when the animals lose most of their size on tables that are turned sideways and expert individuals come in to wield the clippers. Then a fur ball of an animal becomes a skinny creature that walks around embarrassingly naked until the coat begins to return. The fiber is sent for processing and becomes amazing scarves, sweaters, capes, and other fashion garments to delight the wearer and foil the cold. There are shows where the superior fiber-producing alpacas are selected and prized.

However, the happy Wookie world does have drama, and one Sunday a birthing mother alpaca provided more than enough. Hayley had noticed a very pregnant and overdue alpaca showing signs of discomfort, and for too long nothing seemed to be happening except growing distress. The vet was not available on this Sunday afternoon, and Hayley determined to do what she could to help. Usually birthing problems involve the cria turning the wrong way inside the mother and foiling the usual easy exit. When the alpaca lay down, Hayley knew she would need to go into action.

Birthing help requires a mild sedative for the mother alpaca, gloves and lubricant for the helper to go in and set things straight, and a person to hold the mother's head. The only person available to assist was Hayley's brother, Andrew, who luckily happened by, unfortunately dressed for town. Here was a classic clash of Valley life. There is a sophistication and gentile side to the Valley that sometimes clashes with the country realities, and on this beautiful afternoon, Andrew was caught in the middle.

The spruced-up brother had recently come off a reality show, "The Bachelor," where the format involved getting to know twenty-five gorgeous young ladies and, hopefully, selecting one as a potential bride. In Andrew's case, the show did not actually work out into a marriage, but everyone had a great time. He now had a very special steady who the family adored named Ivana, unconnected to the "Bachelor" production, whom he was courting. On the day in question, he was still full of show business and fashion, but he was needed now for alpaca assistance, and like a good brother, he responded to help his sister.

Soon, the mother alpaca, slightly zonked but very anxious, was lying with her drooling head held in Andrew's lap while Hayley began to explore the positioning of the unborn baby. Andrew had been brought up on a cattle ranch where he had seen a lot, but the sight of his sister reaching inside an alpaca up to her elbow was a source of astonishment.

Andrew brought out his cell phone, and he thought he might as well share the moment with Ivana, so he rang her up in Los Angeles, and the conversation went something like this: "Ivana! You won't believe this! I am sitting in a pasture holding a furry alpaca head. She is slobbering on my new clothes and making groaning noises. Hayley is reaching into her backside and I'm not sure whether a baby is coming out or Hayley is going in! There are a couple of dozen of these creatures standing in a circle staring at us. Hayley is never getting cleaned up from this one. It's disgusting!"

The narrative continued in graphic detail while Hayley carefully repositioned the baby, pulling the front legs from under and placing them in the lead to channel out first. She could feel the contractions and, with the unborn in place, she removed her arm and watched with satisfaction as the cria's feet and then legs appeared. Finally the head followed and the rest just slipped out. Andrew continued on the phone to the nonplussed Ivana, describing blow by messy blow with graphic detail.

Soon the baby was standing and breathing with vigorous new life. The released mother licked the baby as though this was just another day in the country. The damp cria unsteadily began to nuzzle the mother, looking for milk. The other alpacas in the field went about their business. All was right with the Wookie world and Hayley beamed with pride of farmyard accomplishment.

Fortunately for our family, Ivana survived Andrew's lurid descriptions of life in the country and was not put off by the messy drama. The two are now happily married, and have produced a young of their own with considerably less difficulty than the alpaca birth.

· 🐾 ·

Dog Day

THE FAVORITE Valley dog day is the annual "Tails on the Trail" Dog Walk fundraiser of the Santa Ynez Valley Humane Society. Dogs look forward to walking their people and meeting their previous acquaintances and making new friends during a splendid tramp across Valley trails on this best of all days of the year.

The Valley Humane Society has been a caring and resourceful animal shelter since the late seventies, with a strong volunteer base and the generous support of the Valley. The worthy organization has found homes and shelter for countless dog and cat friends as well as supplying clinic services and temporary boarding for otherwise lonely domestic pets.

The "Tails on the Trail" is a major fundraiser and publicity opportunity for this valuable resource, and although people do most of the organizing, it is the dogs that have most of the fun.

Dogs enjoy being cleaned up for the occasion, and people are also well turned out, hopefully in proper walking attire. Occasionally some dogs are embarrassed by their people's preparation and outfits, but that does not spoil their fun. The arrival is important, with the most status given to the dogs lucky enough to arrive in the back of an open pickup. The large back space of SUV wagons are next best, and those arriving only in the back seat of a small car will crouch to avoid being seen by their dog friends. After parking, dogs attach their people to leashes and prance to the waiting tents.

"Hello!" says the Lab to an old Poodle friend. "You have gained some weight, and your person doesn't look so fit either."

"Well," replies the Poodle, "we live well and don't have the anxieties of you hunting types! And who is the new chick over there?" Both Labrador and Poodle haul their people over to a cute mixed-breed that has just come off heat. "Hello, doll! I see you are taking your people for a walk. Let's go along together!"

"That lounge lush wouldn't know a ranch road from a front hall," growls the Lab. "Come with me and I'll really show you the country."

The dog conversations continue on as old and new canine friends form up for the trail walk. They organize their people, ensuring that they have registered and made their contribution to the Society, then set off down the path. A prize bag is given to each dog-and-person pair, containing treats, chews, balls, tugs, and other goodies. Some of the good dogs think there should really be something in the bag for people.

A group of dogs with their people congregate at the beginning of the trail. "What is that terrible shampoo you have?" exclaims a large dog to a well-groomed fur-ball dog.

"At least it's better than some dogs I could name, who never wash!" yips back the small dog.

"If you give me that look again, I'll take your head off!" mutters a Bull Terrier type.

"Try it and you'll be overnighting at the vet's," growls the Boxer.

The Shepherd dog quickly tries to organize the slow starting group, saying, "Let's get on the trail or our people will be all tangled in their leashes!" Fortunately the group moves on before any real trouble starts.

Soon the country road is covered with excited dogs leading their owners to enjoy beautiful Valley day. Dogs and people meet each other going back and forth on the road with friendly greetings.

A huge Irish Wolfhound passes a prancing Maltese and remarks as the two stop to observe the people, "That Irish Setter's person looks totally out of shape. I am surprised such a fine dog even brings him out!"

"Yes," replies the Maltese, "some dogs just don't give their people enough exercise."

"But look at that gorgeous female person walking the Pug. I can see she is something from the way the male people are looking at her!" observes the Wolfhound.

"She looks a handsome type to me," replies the Maltese. "I am surprised the male people aren't sniffing her!"

The two continue on their separate ways, with their happy owners walking along contentedly on their leads. The day continues with dog-and-people combinations happily walking the country road. A fat old Labrador lies down panting under the shade of a tree. "Come on!" worries the Shepherd dog. "This is no time to be collapsing with your person on the line!"

"I can only get this person out once a week, if that!" replies the Lab. "Try walking these hills without any exercise and see how you like it!"

Soon the dogs and people begin to return and congregate for the barbecue lunch and dog-and-people socializing. Everyone is pleased to be supporting the Animal Shelter and having a great day in the country while doing so.

A venerable Great Dane settles down happily in the shade of a table and observes to a German Shepherd, " This has been a great outing, but I am content to lie here with my people and dine on scraps."

"I agree," replied a dozy Sheep Dog. "My people have had a good day. By the way, the Animal Shelter is a good cause, but I am just as happy never to see the inside of the place!"

· 🦊 ·

Show Business

IN THE early eighties, a film producer needed a formal fox hunt as background scenes for an upcoming film. At the time, I was managing the hounds for the Santa Ynez Valley Hunt, a club of horse people who enjoyed careening over the countryside chasing a pack of hounds who were following a scent laid down to inspire the runners. In all respects, the Hunt replicated a formal and traditional fox hunt, with hounds, Huntsman, and Hunt Master. The riders wore formal attire, with many wearing the traditional scarlet coats. We were a very photogenic group when fit and shiny horses were included. The only thing missing was the fox, an animal rare in this country, as they tend to be urbanites enjoying trash cans rather than braving the wild country and the dangers of coyotes and bobcats.

When the film idea was offered, the Hunt membership was happy with the revenue offered to the club and keen to enjoy the fun of dressing up and making a film. We made our deal and the film company showed up with cameras, actors, and crew ready for action early one Saturday, a beautiful Valley day. The Hunt members showed up as agreed, as well as my son Andrew, then twelve, and daughter Polly, who had been hired for a special role, riding a sidesaddle.

The first scenes took place in the barnyard, without hounds, with the actors doing dialogue in the foreground and the Hunt members providing the background. There was some maneuvering among the club extras for the spotlight, but we did make a great scene, with well-groomed horses, clean tack, and of course the formal black and scarlet hunt outfits that grace thousands of old English hunting scene paintings. The fun began when we went out on the ranch for action scenes.

The director set up a sequence where we would come at a gallop over a small hill, with the hounds in the lead and everyone scrambling after. Months later, I had a chance to see the film, which was not destined for any Academy Award, but this was the best scene of all. No actors took part in the riding action scenes, but next there were some dialogue set-ups with stationary mounted actors speaking and Hunt members trotting in the background. It was a basic boy-meets-girl film, with a gorgeous chick as a high society Hunt member and a handsome working-class guy trying to woo the heroine. The girl was riding in the Hunt and the boy was tagging along behind on a dirt bike and ultimately destined to rescue his love when she had a runaway horse and had fallen off. Our full Hunt scenes were at first only background, but then the director wanted a few of us for action situations.

The ambitious film called for sidesaddle riding, and there was a stunt performer to take a jump in sidesaddle, and additionally they hired our daughter Polly to do the same. The stunt rider sailed the fence, actually faking the position and was riding astride. Polly had schooled an honest jumper named Chapstick (for his white blaze on the nose) and made a clean jump using an antique sidesaddle with both legs on one side, which is no easy challenge. The Hunt members, watching from the background, cheered Polly. Later, most Hunt members and hounds were excused, while Polly and a few of us were asked to stay on for background.

The crew set up a fake creek that the heroine, and again Polly, were to hop and then visit together about her boyfriend. Polly loved the speaking part. Chapstick was a hero with the sidesaddle, but did not like the fake creek. I was enlisted to solve the problem by popping my Hunt whip behind Chapstick as the two came by, and the horse sailed the creek with a good deal of air to spare. The director and crew had plenty of footage and were very happy with the day, but needed one more shot, where the horse and heroine runs away and the dirt bike hero goes to the rescue.

First they established the scenario by having some of us gallop in front of the camera at a fast pace. Next was the big scene where the heroine, actually played by the stunt rider, runs down the road with dirt bike and hero chasing. Those of us riding background stood in a group watching the action. Twelve-year-old Andrew was riding a polo pony named Sergeant Pepper, who was a sensible horse but a little too high strung. Young Andrew was doing all right, but the Sergeant was getting antsy. When the big scene came, the first horse ran, then the dirt bike, then a motorcycle with camera and cameraman riding behind.

For some dumb reason, Sergeant Pepper decided that it was his turn to run off too, and he took the bit in his teeth and flew off with Andrew down the road, chasing the riders and motorbikes out of sight around a bend. This was one of life's low moments because, although I was well mounted, I could not chase a runaway horse without risking making the situation worse. Actors on horse and motorbike, and camera crew on motorcycle, were down the road; but with a runaway behind, this scene had to end with trouble. The bad news was radioed back to the waiting crew that there had been an accident around the bend.

I arrived to see the Sergeant being held without Andrew

on board and the crew staring down the side of the road. When Andrew galloped behind the motorbikes at a runaway pace, the procession had gone around a bend and then stopped. The Sergeant, faced with a horse and two stationary bikes across the road, came to a screeching stop and Andrew flew off and down the side of a steep bank. Fortunately it was a soft hillside, which saved Andrew's hide but also almost asphyxiated him from being buried in the loose, dry dirt. The stunt rider was the first to Andrew, down about thirty feet from the road, and he quickly wiped the dirt out of his mouth and face. A hasty check revealed a gasping son, but one without major injury beside a few scrapes.

We made our way up to the road and the concerned film crew, who quickly cleaned his scrapes with a first aid kit they had brought. When all was calmed down, we did the only thing that should be done in those circumstances. Andrew was plenty shaken, but, to avoid any further dread of riding, a fallen rider needs to get back on and ride immediately. With some complaining, he did so, and we both rode back peacefully the two or three miles to the barn. We had had enough of show business for one day.

Ultimately the Hunt Club proudly viewed the production and admired the background scenes. We had worked hard for our fee, but after seeing the film, were glad we had received the money up front, because the film was not going anywhere with success. Show business was a great experience for us as it has been for many Valley characters who have played extras in the films produced in our country.

· *F* ·

Narc Nose

WHILE LIVING in Chicago in the early '90s, a young animal lover named Charlene Hiatt became determined to develop a partnership with an honest working dog and researched the possibilities for such a career. She discovered that a search-and-rescue partnership with a German Shepherd would be a very satisfying avocation that would allow her a public service opportunity and a dog-partner friendship. After researching the possibilities, she discovered how to pursue her dream and the means of finding a suitable partner.

She was introduced to a well-bred and talented female German Shepherd and paid eight hundred dollars, a goodly sum to her in those days, for the second male pick of the dog's future important litter. Top American German Shepherds, such as the great Mystique included in this book, are bred for the show ring, while the European German Shepherds are bred for work. They have the stature, disposition, and talent to be police dogs, guard dogs, and search dogs of great usefulness. Charlene had learned that this very talented female was to be sent to Germany to be bred and would then return with the progeny to Chicago. The honeymoon trip and family were accomplished, and a loving and talented puppy to be named Fritz came into Charlene's life.

From the beginning, Fritz was brought up to be a working dog. He was not to be spoiled with frivolous play or affection, but was taught the basics of command and control from an early age. He grew up with the very desirable attributes of a husky conformation, with broad shoulders and high stern quarters that signify the German working breed. He had a keen mind and, best of all, an extremely long and talented nose. The tradition with these dogs is to teach them commands in German, hence "Zit" meant sit down, "Plotz" meant lie down, "Nein" was no, and so on. It soon became evident that Charlene had a willing and talented dog friend who was ready for serious training.

The pair entered a six-month training course that was attended mostly by police officer-and-dog partnerships, but Fritz prospered as a private citizen in school. The serious police management soon realized that Charlene was for real in her desire to grow a search-and-rescue partnership. The training for this non-police partnership also involved Charlene with map reading, topographic skills, and radio training and other law enforcement–type skills. Fritz progressed from a

Basic Dog degree to an Advanced degree and finally won the approbation of the class teachers with the coveted "CGC," or Certified Good Citizen dog recognition diploma.

During the training, it had become evident that Fritz possessed a superior ability to sniff out whatever he was trained to find, and with really good sniffers, that leads to a career in discovering contraband drugs. Soon the search-and-rescue pair began to train for illegal substance search. Dogs discern scents differently and much better than people. People will smell a stew and receive the impression of stew. Dogs will smell a stew and, separately and individually, discern meat, carrots, onions, potatoes, seasoning, and more in a clear catalogue of individual scents. Some people believe they can hide contraband from drug-sniffing dogs by mixing in coffee or hiding the stash behind a gasoline can. A drug-sniffing dog will take a sniff and think, "Well, well," or in Fritz's case, "Ahh Szo! Here is both coffee and marijuana," or "Here is gasoline and methamphetamine."

Fritz soon became a master at sniffing out whatever illegal substance was hidden, and the law enforcement types who ran the class were impressed. The dog would find and then alert, which means he would sit, staring at the source of the offending odor, until released by his handler, Charlene, who then brought out the tug. Fritz was crazy about playing tug with an old braided piece of material. He enjoyed the game of tug-of-war, but also realized in his doggy mind that the tug was brought out when he had performed well, and only then, and Fritz was born to please. Tug was never played frivolously, but only as a reward for doing well and thus became the recognition of a job of searching for drugs well done.

Normally, a talented dog like Fritz would have an important career in partnership with a law enforcement officer, but Charlene had no desire to do police work. A crucial distinction between a sworn officer and a citizen in a drug-dog partnership is that a private person is not compelled to report an illegal substance, as is required of an officer of the law. The two, Charlene and Fritz, formed a significant and unusual partnership for a confidential, private drug consultancy.

About this time, the pair moved from Chicago to the Santa Ynez Valley.

Fritz loved living in the Valley without the snow and wind of Chicago, and the two soon began to grow a reputation as the only private drug-sniffing team in the County. As an example, if a family suspects that one member, usually of the teenage variety, is engaging in an illegal recreational pharmaceutical activity, they might want to discover the facts before any confrontation. Understandably, the family would not bring in the authorities, so a confidential visit by a private drug-sniffing dog would be able to establish the facts without unwanted publicity. Charlene and Fritz's fee was one hundred dollars an hour plus transportation, which was good compensation for Charlene, while Fritz continued to work for tug recognition.

A typical scene might include a visit to a home in the quiet of an afternoon. Fritz would be given the run of the house and would often find and alert, sitting and staring at a piece of furniture or other hiding place. A quick search would usually reveal a stash of contraband and the family would know the facts. Charlene never touched or acknowledged the illegal substance or performed any service or gave any advice to the family. A few minutes' reward of tug with Fritz and the pair would be on their way, having performed the service of discovery. Many families in the Valley have been significantly helped and have avoided unfortunate consequences, thanks to this pair.

Public schools use only the services of law enforcement drug teams, but private schools have called in Charlene and Fritz a number of times. The pair had only to pull up in her SUV with the NARC K9 license plates to start a buzz throughout a school and often a hasty discard or exit from the campus. Cars and lockers have revealed contraband knowledge that has significantly helped schools maintain a drug-free campus.

Once, the pair was on a public school campus doing a class demonstration and the principal suggested that Fritz might want to visit a certain boys' bathroom that had been under suspicion. Of course this was completely unofficial, but Fritz did become seriously interested in a certain light fixture above a commode. Charlene quickly left the campus and only heard by rumor that law enforcement had made a hasty visit to the campus also, and an ongoing drug problem was cleaned up in short order.

Fritz finally retired from his intense service job due to an older dog's illness. He is more relaxed now, and enjoys a game of tug without working for the reward. But needless to say, if anyone visits Charlene and Fritz with a stash of contraband substance, they should be prepared for exposure.

· ℱ ·

Diplomatic Auction

THE VALLEY lost a center of our Western heritage when the Santa Ynez Sales Yard went out of business in the late eighties. Prior to closing, every Thursday ranchers, cowboys, and a few cattle dealers and buyers would congregate at the yard, about three miles north of Buellton on Jonata Park Road, just off 101, to buy and sell cattle. The pens and sale house can still be seen from 101, and on close inspection, one can imagine the ranch drama that took place there as the prices fluctuated and herd operators tried to make a living. Charlie Hollister ran the show and acted as auctioneer if he could not find a professional to call the bidding. In the yard's heyday a few hundred cattle from all over the tri-county area would be run through the sale on a Thursday.

The routine would go something like this. Starting a day or two before the sale, ranchers in trucks and goosenecks would deliver their stock to be sold to the receiving chutes. The County Brand Inspector would examine the brands to verify ownership, as cattle could not be moved without a shipping form which required the brand for identification. The delivered cattle would be weighed in and sorted into saleable lots and held in pens, with a number sign pasted on the hips for identification. On the day of the sale, buyers would walk the pens and plan their purchases. At noon the auction would open with the first lot run out of the pens, down the lanes, and through the heavy gate into a well-fenced and brightly lit sales shed. The buyers would sit on bleachers, protected from the cattle by cables, with the auctioneer and recorders stationed in a raised vantage booth sitting over the cattle opposite the buyers. Two experienced hands with whips would dodge in and out from behind protected stands on each side of the auctioneer box, like the barriers in a bullfighting ring, to circulate the cattle for complete inspection by the buyers and then move the lot out when the sale was made. Cattle, and especially the odd bull, can kick and butt, and these stock movers needed to be quick and cow-smart.

The ranchers, cowboys, and family members would sit on bleachers on the buyers' side, behind the cable fence, and make their calculations during the bidding action. The big buyers would be from feed lots and packing houses, and those people usually sat in the front rows. Behind, small-lot buyers wanting to stock ranches would compete for good young cattle at a bargain price, hoping to make money on weight gain over the months ahead and, with luck, a price increase when the grown cattle returned to the auction yard. Sometimes the drama would mount for a rancher who had brought his crop of yearlings and hoped to make a dollar or two. The bidding was usually spirited, with the auctioneer running on in a barely decipherable but lively and loud monotone, trying to be heard over the noise of the cowboys yelling and cattle bleating, until the sale was made. The cattle were then driven out onto a scale and the weight of the lot was posted on a large lighted sign. Price and weight were sent immediately to the office for a cash settlement, and the ranch sellers always returned home with a check in their jeans.

Many western types showed up without cattle just to hear the gossip and watch the action, and there were always friends hanging around outside, where sometimes there was a steak sandwich barbecue. In the days after the auction, shipping and receiving action usually continued in the yard pens on through the week.

One day, because I was one of the locals who supported Vandenberg Air Force Base, the Commanding Officer enlisted me to help entertain a group of Russian military rocket people. This was in the early days of Glasnost, when the Russians and Americans were learning to be friends instead of mortal enemies. The fact that these visitors had spent their lives wanting to kill us was forgotten in the new spirit of friendship, reinforced with mutual military inspections. This high-ranking Russian delegation expressed an interest in seeing some real American people after the formalities. I suggested to the Air Force people that the best way to do this was to observe some real cowboys who hung around the auction yard. The Vandenberg authorities cleared the idea and the advance diplomatic and security people scoped the yard and set the schedule. Charlie Hollister spread the word that our Valley sales yard was about to play an important part in international relations. The locals spat tobacco and looked forward to the day.

When the big moment came, we got the call that the large party of Air Force and Russians was about an hour late, but very much looking forward to experiencing the real West. Unfortunately, when the visitors were now scheduled to show, the cattle would have run out and the sale concluded. But a hasty word went out to all the characters assembled for the day's auction. The honor of the Santa Ynez Valley Sales Yard was on the line.

When the Russians arrived with the nervous Air Force brass, they found the sale in full swing. Ranchers and cowboys were engaged in spirited bidding over large lots of cattle run down the lanes and into the sale space with horses running and auctioneer yelling. The Russian generals sat in the bleachers and realized all the fantasies they had imagined from Western films shown on television reruns during the long Siberian winters. Here indeed was the Wild West before their eyes! A good time was had by all, and it was impossible to tell that the whole thing was faked. The yard had come through and the military was happy.

It is sad that the modern cattle business no longer supports the local sales and the characters that frequented the yard, but it was a great Valley center for Western heritage in its day.

·*F*·

Fundraiser Penning

IT SEEMS that St. Mark's Church has always been a central part of Los Olivos, but there was a time when the present church location was a vacant lot. In those days, the congregation was busily planning to build a new church and in a daft moment some of us from the congregation decided to put on a team penning as a fundraiser and use the future church site as our pen. It was heavy-duty work, but we didn't know any better, and the results were very gratifying.

In those days, team penning was new to the Valley and mostly a social and family event with everyone competing. Recently, it has become a more professional, high-stakes sport, but then, because it was amateur and friendly, anyone could play.

A team penning arena is the size of a horseshow ring with space inside for a smaller pen placed at one end large enough to hold three steers. For the event, a herd of thirty cattle with numbers stuck on their hips in sets of three will be put into the arena. There will be three ones, three twos, and so on. The cattle will be at one end and the pen at the other, where the team of three riders will enter. The idea is for the riders to pen their three steers of a given number as quickly as possible.

To begin, a team of three nervous riders would enter the arena and stare down thirty spooked steers. When they ride across the timer's line, the three teammates must try to separate three animals of the same number, herd those steers and only those steers into the small pen, and try to do this in the best time possible, with a maximum of three minutes allowed. Cowboys and cow horses can do this if luck is going their way, but dressage riders and show-horse types are usually frustrated. The format proved to be a great potential for our fundraiser with a large crowd, many entry fees, and generous spectator contributions.

On a spring Friday of the St. Mark's fundraiser weekend we began hauling the heavy green Powder River fence pieces to construct the pen and corrals. Many bruised fingers and sore backs later, we had our arena and pens constructed on the site of the present church. On early Saturday we ran the cattle down the corrals of the two ranches we had talked into lending us cattle for the event. We stuck patches high on the rumps, showing three ones, three twos, and so forth until we had four sets of thirty cattle ready to be penned by the contestants. This is hot, dusty work and required some Episcopalian beer for the volunteers to survive.

On the big Sunday, we all went to early service and then began hauling cattle. The cattle were not as enthusiastic as the volunteers, and required persuasion to enter the new pens. The entire congregation pitched in. The Altar Guild set up coffee and doughnuts and the Men's Club fired up a barbecue. Hay bales and chairs were set up for bleachers and a sound system that could only be tamed by our fourteen-year-old son, Adam, came to life.

We were ready, and a Valley horse crowd of all persuasions began to arrive. Jack Algeo, our own announcer, tried to explain the rules to the spectators, and Gordon Wooden set up the cattle. Our youngest and cutest were designated ticket sellers and worked the growing crowd unmercifully. A large number of locals had turned out as well as many uncertain contestants, and we made the most of it.

Dozens of riders ringed the arena waiting their turn behind the spectators, who were sitting on anything they could find. Greetings, challenges, and insults abounded. Coffee and Danish pastries circulated and barbecue aromas filled the air. And then the first teams began.

The initial competition was set up for only members of the same family to compete. Teams of three family members tried enthusiastically to pen, but only the fourth team was successful, to the roaring approval of the crowd. The day was off to a great start and the competition became more intense with other designated teams. People on Arabian or Thorough-bred horses that had never seen a cow got into the act. Heavy side bets changed hands. Skill was cheered and incompetence was cheered. At one moment, our indomitable rector, Father Stacey, the number-one popular entry, rode a borrowed horse with total crowd support if not glory. Chuck demonstrated that spiritual sheep and goats are more in his line than steers, and his team was cheered but won no prize, even with the help of the timekeeper. Episcopalians even cheered when a prominent group of Presbyterians won the amateur competition.

Finally, the serious open class began, and this time for real, with trophies up for the winners. A team of top hand cowboys came in looking like they meant business. The crowd hushed as they crossed the line and sent a first-class cutter into the herd who quickly separated one, then two, and then three steers of the designated number, while the other two held the herd back. The team quickly bunched the sep-arated three and drove them toward and then into the pen, raising their hats as the signal for time when the steers were held in. "One minute, five seconds!" shouted the announcer. Three cowboys from the Sisquoc Ranch had come to town and won silver buckles!

Competitors and crowd continued to enjoy barbecue and refreshments and a good time was had by all. The treasurer counted out a few thousand dollars, Chuck Stacey shouted "Hallelujah!" and as the horse trailers and spectators began to depart we knew we had a great success. Of course there were cattle to be returned, fences to be put away, and all the cleanup from a great day to be accomplished. That night, af-ter the take-down, volunteers repaired happily to the Mattei's Tavern bar.

Today that same team penning space is sacred church ter-ritory, but we all know that the site is extra blessed by the hooves of horses and steers that contributed their part for our St. Mark's Church!

· 𝓕 ·

Cat Therapy

A CASUAL CONVERSATION with a psychiatrist at a cocktail party produced a most splendid animal saga and a productive life lesson. The good doctor, who of course must remain nameless to protect client relationships, lives in the Valley and practices in Santa Barbara, where apparently there is no lack of psychological patients. Many of his clients have been brought from trouble to healthy, mentally balanced lives through the skills of the doctor, but also through the naturally therapeutic skills of cats. Here's how it works.

Years ago the doctor concluded that many of his patients were not suffering from true mental illness, but rather a loneliness and boredom which led to some of the benign but real symptoms that brought his clients to the psychiatrist's couch. All of us need to give and receive love, and sometimes circumstances are such that there is more of the giving than the receiving. Here, often, cats can redress the balance. If the circumstances are right, the doctor explained, all the science in the world cannot match a good cat at work.

In the course of treatment, when cat therapy is called for, the process begins with the insertion into the therapy session of one of the mother cats the doctor maintains in his office area. While the patient rambles on during the costly hour, the doctor will begin with the cat in his own lap, chummed with snacks and stroking. The patient will then begin to notice and accept the cat as a routine fixture. Next, the doctor subtly transfers the snack distribution to the patient, and almost always, there will be a growing bond between cat and patient during the sessions and hopefully the cat will look forward to the warm and friendly conversations and dozing in the patient's lap.

The next ingredient in the psychodrama is for the mother cat to do the natural thing and go out on the night streets to meet a romantic and enthusiastic tomcat and become pregnant. In the ensuing weeks, if all is going well, the patient, on the road to recovery, will be invested and involved with the blessed cat family event. Soon, as nature takes its course, there will be a litter in the doctor's office to supplement the application of cat magic healing power.

Meanwhile, the normal scientific treatment has progressed and the patient, growing in mental health and stability, begins to approach the day when office sessions will no longer be needed. Of course none of us are really mentally healthy and we all have our quirks and needs, and here the doctor recognizes the requirement for ongoing therapy outside the formal sessions. To bring this about, the doctor contrives to transfer the patient's affections from the mother cat to an adorable and adoring kitten. This is usually accomplished easily, given the nature of kittens. Now a growing patient psychiatric wellbeing will be supplemented and strengthened by a growing affection of a kitten.

The final act is played out with the patient saying good-byes to doctor and staff and departing to a healthy and mentally sound life in the outside world. And, of course, the patient departs with a basket with ribbons, cat toys and adopted kitten. The inscrutable and mysterious cat mind of the kitten will now accept the responsibility of ongoing cat love and therapy with the adopted patient.

"That's wonderful," I exclaimed. "Does it happen often?"

"Well. Truthfully, no," the doctor replied, "but when it does, it is a joy!"

· *F* ·

Poacher Roachers

VALLEY COUNTRY is, by and large, a law-abiding area and this is partly because locals take care of their property and each other and also because of a first-class sheriff's organization. But rural crime can happen. There are rare occurrences of cattle rustling on Valley ranches and, more often, bandit hunters will poach game from the open hills, sometimes doing the butchering at the scene of the crime.

Property owners tend to be protective of their land, and

Santa Rosa Road rancher Johnny Wiester takes his steward-
ship very seriously, particularly as the ranch is on both sides
of the road. The Wiester Ranch is off limits to all hunting,
even during game season, and sanctuary causes game to gravitate to this safe haven. A problem for deer seeking refuge is
that Johnny occasionally stalks the unsuspecting buck with
bow and arrow himself, rarely with any success. The deer
seem to know this and have learned to live with the odd arrow plunked through the brush in their general direction. Far
better that than the boom of a high-powered rifle directed by
a telescopic sight.

One year during deer season a particularly impressive
buck had taken up residence on the ranch and learned to
dodge arrows. The huge buck lived in watchful alert, showing
himself now and again to the satisfaction of the ranch people
who had granted the asylum. But trouble loomed.

John Wiester had become aware of a stranger's pickup
truck that regularly cruised Santa Rosa Road. The vehicle exuded suspicious vibrations, and a canvas tarp was observed
in the truck bed, surely waiting to cover the carcass of an
illegally taken buck. The truck drove slowly up and down
the road and daily became more suspicious. Finally, one night
the truck was left abandoned at the side of the road while the
driver was obviously on the ranch, hunting.

The sheriff was called and a deputy soon showed up to
stake out the parked truck. Cops and ranchers sat in the darkness and awaited developments in watchful anticipation.
Johnny had enlisted the help of a neighbor and friend, Bill
Suter, who felt strongly about game protection and his past
experience in Viet Nam gave him the ability and penchant
for action. The two friends became bored and frustrated with
watching an empty truck while the poachers were hunting illegally on the ranch, and decided to stalk the evildoers where
they knew the trails and topography better than the trespassers. They knew where the buck hung out and where the
poachers were likely to be. Of course they were unarmed and
in the dark, but right was on their side.

After a frustrating couple of hours of night stalking, they

returned to discover that the stakeout deputy had been called away. The two sat in their car wondering what to do with the poacher's truck behind them on the road. And just then the offending pickup, with trespassers returned, started up and sped by Johnny and Bill. They had just a moment in the head-lights to see something bulging up under the tarp. They must have killed the buck!

The chase was on down Santa Rosa Road and into Buellton. Both pulled up at the Highway 246 stoplight, and Bill rolled down his window and yelled, "Stop, we want to talk to you!" The answer was a squeal of tires as the pickup sped past Anderson's and onto Highway 101.

Buellton has a Highway Patrol station and Patrolmen and Sheriff's deputies drink their coffee at Anderson's and Ellen's Pancake House. Any local knows the Avenue of the Flags is a bad place to race cars. Soon the law was in hot pursuit and the airwaves broadcast the chase to all available law enforce-ment. Poachers were alarmed and ranchers were angry and righteous, and the pursuit ran up 101 at over ninety miles an hour.

The Highway Patrol stopped Johnny and Bill first and started to write a very big ticket. Johnny blurted out the story to an unimpressed Patrolman, but soon the airwaves crackled with an even bigger story. The Patrolman led Johnny and Bill to make an identification at an important arrest scene up the road with many red lights, weapons drawn and "poachers" handcuffed.

When a second patrolman had stopped the "poacher's" truck, a loose tarp had revealed a huge load of freshly cut mar-ijuana. It became clear that illegal pot was growing on some obscure corner of Johnny's ranch, unseen by the rancher. This can be a very embarrassing situation for a rancher, and some-times dangerous if discovered. There are often places on large cattle ranches that are rarely visited by normal operations, and the pot growers can sneak in and take advantage to boot-leg a more valuable crop than cattle. In this case, the "poach-ers" had been stalking the patch and snuck in to steal the fruits of someone else's labors. The big buck was an innocent bystander to a much more earnest drama. When followed, the "poachers" thought the original growers had spotted them, and the outcome could have been serious. But the arrest was serious too.

The Highway Patrol let our heroes, Johnny and Bill, off the ticket with mutual congratulations. The next day, there were sheriffs and helicopters all over the Wiester Ranch searching out and destroying the remaining pot plants, but the growers were never discovered. The buck took alarm in the confusion and was never seen again. Quiet returned to the Santa Rosa Road ranching community and, hopefully, all the growers and poachers alike have returned to Merced County where they belong.

· *F* ·

Protein Prayers

A RT KNIGHT is a much appreciated Valley character who has been involved with 4-H programs as a judge and leader for many years. Before retiring, he was a Deputy Sheriff who ran the rural crimes division of the Santa Ynez Valley. He was surely hell on wheels for the bad guys, but for the rest of us, with only a little on our conscience, he seemed more like a little league coach or the usher at Sunday service. A few years ago he undertook to help his daughter, Darcy, win a 4-H prize at the County Fair.

The basic 4-H program is to teach young potential farmers about husbandry through the experience of raising and bringing an animal to market. After the type of animal is chosen and selected, management of the project will be supervised in 4-H Clubs, and an adult team leader will work with the young person for a season, culminating in the County Fair judging and sale. The animal will be spoiled with unlimited feed, supplemented with oatmeal, M&Ms, Coca-Cola, or anything else a hungry rabbit, chicken, sheep, pig, or steer might wish. The animal prospects are bathed, trained, and pampered in the care of the 4-H member and family. The finished animal will be judged at the Fair, perhaps becoming a class winner or, best of all, a grand champion that will fetch a few thousand dollars and a highly coveted prize. The first challenge is to choose an animal that has the potential to grow up great.

Like a good dad, Art consulted with his daughter, and they decided to raise a steer. The first challenge was to find a good prospect and Art lined up his appointment to select a steer that would shine. He drove out to the San Lucas Ranch near Lake Cachuma to deal with the foreman, Gene "Squeaky" Matthews, who was a first-class cowboy, a character, and a good judge of cattle potential. The two inspected the herd and chose a strapping Santa Gertrudis youngster with great potential. They chose well on the physical side, but slipped up on the character part.

The Alisal Ranch, near the Knight property, had bought a load of steers from the San Lucas to winter on the ranch, and Art arranged to have the chosen steer included in the shipment, then to be transferred to the nearby Knight home pasture. When the day came, Art drove his trailer to the Alisal and the cowboys cut out the steer from the herd and loaded the animal. The Alisal hands seemed to be a little amused with Art's choice, but he paid no attention and trailered the steer for delivery to Darcy.

The Knight family home has a pleasant green pasture made to order for the new boarder and all was ready to spoil him with buckets of mash, sudsy scrubbings, pleasant walks, and love and affection. Steers are benign animals and will sell out easily for a bucket of mash and a back-scratching wash. However, on the next morning, this new boarder took one look at Darcy approaching with bucket and brush and promptly chased her up a tree that fortunately stood with a low-hanging branch in the paddock. Art rescued his daughter by shaking a two-by-four at the steer and wondered what he had brought home.

The next day a group of fellow 4-H members mustered their support to back up Darcy and tame the brute. The idea was to settle the steer by stages without risking life and limb.

The angry Santa Gertrudis soon had the entire frightened 4-H team up the same handy tree.

Art was losing patience, but remained determined that this steer would grow into champion material. One of the requirements of the program is to lead the steer by halter in and out of the judging ring. To help his daughter, Art somehow managed to get a halter on, but the leading part required the two-by-four by way of persuasion and protection. The steer was going nowhere Art had in mind and seemed to be getting the upper hand in the training program. After a few days of standoff, the Knight family understood they had a real problem.

Art's slow burn was heating up with the unaccustomed lack of respect from the steer. Finally, he took a bucket of grain, and the two-by-four, and also strapped on his service pistol to have a visit with the animal. There was something in Art's attitude that impressed the steer, and instead of snuffling and charging, he backed off, took a run at the fence, and hopped into the neighbor's yard. Here the saga took on a new twist.

Their next-door neighbor, Valerie Scott, was a devout lady, given to prayer and good works. She cultivated her garden to raise food to feed the Mission effort for the homeless. Valerie thanked the good Lord for this garden bounty, but had added prayers for more and better nourishment for her homeless friends. She was a country person who knew how to raise food, and knew the value of a steer converted to protein for the table of the needy. Now, the animal landed in her yard looked to her like ancient manna dropped from heaven for the children of Israel.

Art had always liked and admired his neighbor for her good works, which he considered as he knocked on her door to apologize for the bandit steer in her backyard. Of course, by then he was thoroughly fed up with the animal and had given up on any 4-H possibilities. As they visited and discussed the situation it soon became evident that the neighbor's prayers had been answered. In those days there were a number of butchers that would visit a ranch to shoot and cut up a steer for the family table. License to kill was granted by Art forthwith, the butcher called, and the steer soon made a welcome addition to the Mission homeless shelter fare.

The sequel to the unhappy 4-H steer adventure was a pig project with a new, more friendly animal, who became a happy camper with the Knight household and a very successful participant in the Santa Barbara County Fair.

· *J* ·

Part Four

Ranch Work

Cattle Realities

D R. JIM Shoup is an accomplished veterinarian who specializes in keeping high-end Thoroughbreds healthy, running, and reproducing. He is a friend and took my distress call on a Sunday afternoon, although like most Valley vets, he would respond to any appeal from an animal with a dire problem. On this day in the early 1980's, we had a calving cow unable to get the job done, and in danger of losing her life in the process. None of the ranching-type vets were available and Jim answered the call. "I haven't done any cattle work lately," he said, "but I will do my best." Jim's best was over-the-top best.

The young cow had been in labor for most of the afternoon when we spotted her lying on a hillside. Something had gone wrong, and the calf was not dropping, or even moving, and her pain and danger were evident. Jim showed up in his vet's truck and maneuvered to the quiet but obviously troubled cow lying on our ranch hillside, fortunately close to the road. A quick examination revealed that the birthing had been going on too long without success, and an operation would be necessary to save the cow and, hopefully, the calf. Ranch foreman Bill Reeds had a rope on the cow, but she was never inclined to get off the ground, and quietly submitted to treatment.

Further examination involved Jim putting on a long glove that covered his entire arm and reaching far into the back end of the cow to discover the full extent of the problem. Jim stood up and said, "Brooks, we'll need to do a Caesarian operation, and since you got me out here, I am designating you as my assistant." I swallowed and put on the operating gloves, following the lead of the master surgeon, and joined in with an unsteady hand.

Bill held her head while Jim gave the cow an intravenous knockout and proceeded to set up an outdoor operating room, laying out all his instruments on a sterile cloth. The unconscious cow was scrubbed and shaved and her bulging Black Angus side opened in a precise and masterful operation, layer after layer. Soon a large, but unfortunately dead, calf was exposed and removed, which never would have been born under normal circumstances. Jim's every movement was that of a skilled surgeon who had served so well in the Thoroughbred operating room, but now played out on the open range. He sewed the cow back up with great pride and precision, with some amateur help from me. He stepped back proudly to survey his work as the happier cow slowly regained consciousness and stood up. Otherwise, it was a beautiful sunny, Valley day on the green hillside with the cars speeding along 101, oblivious to the country drama playing out nearby on the ranch.

"I haven't done honest ranch work in a number of years," Jim exclaimed. "It feels good to be back in veterinary school! But I don't think this cow will breed again. What will you do with her?"

"I guess she'll move on to the sales yard," said Bill.

Jim reacted with amazement. "Is that all she can do after I saved her?"

"What did you think, Doc? We'd send her back to the racetrack?"

"Yeah," the good vet replied, "I guess I should have known."

Life isn't always easy on a cattle ranch. To Jim's credit, he billed us as a ranch job and returned to his important Thoroughbred practice.

· ℱ ·

Branding Day

VALLEY CALVES are born into a friendly world and in their early days bloom with amazing health and vigor, thanks to a doting mother cow whose best satisfaction is to turn on the warm, sweet milk taps. Calves gambol and frolic under watchful, broody cow eyes as the mother munches grass. But this bucolic bliss brings growth and entry into the commercial world of ranching.

Cattle management means good health through inoculations and vitamins, the legal requirement of a brand mark, and, for the bull calves, the gelding indignity that turns them into steers. These requirements mean the herd of mother cows and calves must be gathered in from the pastures and in the ranching cycle that is referred to as a branding day. Here is a typical Valley branding, a day in the ranch cycle very important to our ranches.

The branding is planned well in advance, with all the details of medicine and cow management and the crucial question of help. No ranch has enough people to handle the branding day chores on their own, so a number of friends and neighbors are solicited, most of whom have been through the process with the ranch before and most of whom will, in turn, ask for help on their ranches. The process is governed by conventions that are as old as the ranching West.

The day will probably begin well before dawn, with horse trailer rigs arriving at the corrals. Riders will unload the horses, saddle up, and mill around, complaining about the cold morning while the host ranch serves coffee and Danish pastry in the dark. When the important gossip is handled and horses are ready, the assembled company will ride off in the frosty, dark morning. Valley pastures go from two hundred to two thousand acres with hills, trees, brush, arroyos, valleys, and numerous other places for cattle to hide. The number of riders will need to be chosen to match the size of ranch and herd, which in the Valley can go from a token ten or twenty to a few hundred cows and calves. Everyone knows that an experienced cowboy is worth four green hands, but the rancher usually takes what is available and appropriate to invite, hopes for the best, and all ride out together.

The idea is to surround the herd in the early cool before the rising sun warmth causes cattle to brush up in tree shade. The herd is then gently moved off the pasture into the corrals, leaving none behind. If all goes smoothly in the early stage, the riders will be at the far end of the pasture and ready to fan out and round up by first light. Hopefully there will be no morning fog to hide the herd.

The Valley atmosphere has a special air on these mornings, with smells of grass, sea air, horse, leather, cattle, and a unique and indescribably sweet central California aroma that makes life very worthwhile. The horses are ranch fit and might be frisky in the fresh morning air, but ready for the long trek. They have all been this route before and they know they will win and the cattle will lose but it will be a long day. Most horses seem to take an interest in a roundup, although nobody really knows what horses think.

The sun will rise and the beauty of the dawn and the welcome sun rays will warm and light the riders on their separate ways through the back parts of the ranch. Soon most of the riders are quietly on their own and hopefully not lost, as the cowboy net is cast around the pasture and waking cattle. The riders turn toward the desired gathering direction and seek out the animals for a gentle and thorough sweep of the countryside toward the corrals. The idea is to move the herd steadily and smoothly without excitement and absolutely without losing any in the sweep. Here cow sense is very much required, and the wrong move can send a cow or panicked calf the wrong direction or up an impossible hill. One rider might go an hour without seeing a cow and another might take on the responsibility of driving half the herd. Deer and the odd coyote will dash away, wanting no business with the very suspicious-looking people.

After a sometimes long and lonely trek, riders will hear or see others, moving at about the same pace and drawing toward the desired area. If all goes well the herd will gradually and peacefully assemble and the cattle will filter smoothly through the gate, into a holding pasture or corral. There will be a few "Yup, Yups" or "Hyahhs," but very little excitement or rapid movement. Here again, cow sense is at a premium and a wrong move can lose a good part of the herd back to the hills or cause a dashing calf to bolt away from the pack. The form is to exert exactly the right amount of pressure, but not too much, and to continue the momentum, pushing the cows and calves to where they probably do not want to go.

Almost always, the herd ends up in a dusty, milling mass in the corrals with cows and calves mooing and bleating to pair up in the confusion. A relieved cow boss and happy ranch family and friends now tie up their sweaty horses and enjoy a cup of coffee before the work begins again.

Everyone must now know their place for the next round, the cattle sorting. The top hands will do the cutting out to separate cows and calves, the next level of riders will do the herd-holding in the large corral, the next level might work the gates on foot as the cow boss calls where the sorted animals should go. The least experienced or less well-mounted participants and visitors line the fence and watch the action. Soon calves will be separated from the cows and the real work will begin.

One corral will be the branding area, with a fire for branding irons and tables with medicines and instruments laid out. Usually groups of ten calves are brought into this working area where the first teams of designated ropers wait, swinging their loops. A roper will first sling a noose onto the head of a calf and pull the jumping animal into the path of a second roper who will catch the two hind feet. The calves will be held between the header and heeler, ready for treatment.

The ground crew will immediately put the calf on the ground and change the head noose to the two front feet, and the calf will be stretched and ready between the two holding horses. This is a skilled teamwork job coordinating riders, horses, and ground crew and is the pride of the participants, the entertainment of the assembled watchers, and the brief discomfort of the calves who seem to forget the ordeal a moment after release.

The fast and hopefully well-coordinated ground crew will give the calf three or four injections, a brand, ear spray, miscellaneous doctoring as needed, and a rapid operation to change the bull calves to steers, all in seconds. Meanwhile, another team of ropers is catching the next calf and so on until all the young crop of calves are branded. Who ropes and who works the ground crew and who watches is orchestrated by custom and the dictates of the cow boss and very much understood by the assembled company. Skill is praised, mistakes are derisively forgiven, and bad behavior unthinkable. The sun is high now, and the work becomes long and often hot, but team spirit and ranch energy must carry the day. What started out as an adventuresome riding event has now become a long morning of hard work.

Finally the herd will be sorted, branded, doctored, and quiet. The last chore is moving the animals back to pasture, where the happier cattle, having forgotten all the indignities and discomfort of the day, make their way back to the hills. The best moment is watching the departing herd, with all mother cows healthy and mooing for their offspring, pairing up with the gamboling calves, free on the range and at home on the ranch.

Back at the corrals, the horses will be watered, unsaddled, and tied to the trailers, munching on a pack of hay. Their day is over, and the fun has just begun for the assembled company. Form requires a good feed and drink for the volunteer help, and in the Valley, that is always forthcoming. The usual is a tri-tip steak barbecue with beans, bread, and salad. In the old days, beer and the hard stuff supplied the bar, but since the seventies, some excellent Valley wine usually graces the branding day tables. Form calls for invitees to bring a dessert, so the main course will be followed by heaps of prideful sweets.

Cowboys, friends, and neighbors, tired and happy from a long morning job, will join invited spectator guests for a hearty meal at trestle tables. The morning's good and bad moments will be rehashed and enhanced to the joy or sorrow of the participants. Finally goodbyes will be said as the trucks and trailers go down the road. With luck, the host ranch has had a good count of a healthy calf crop and, and if the prices hold, will enjoy some ranching prosperity. As the calves and cows meander back to their Valley hills in a happy, bucolic cattle way, the worn out host ranch will take satisfaction in another annual branding. This routine has been with the West for generations and hopefully will continue in the Valley for generations.

· 𝒥 ·

Outwitting Chickens

THERE IS something about chickens that provides good vibrations. I never met anyone who did not like chickens, except possibly the person who cleans out the cages or is awakened by a rooster at dawn. Chickens strut around making comfortable clucking sounds and scratching and pecking the dirt in a benign and reassuring way. Hens lay eggs that are as appealing as a marble statue or new-fallen snow. There is hardly anything wrong with chickens except their light-weight intelligence, and they do have their different ways.

Shortly after we moved to the Valley, we populated an abandoned hen house on our ranch with a small flock, thinking of fresh eggs. Two dozen white Leghorns moved in and soon were content to produce more eggs than we could eat. They clucked and scratched, pecked at the mash, and pooped into the water dish. All was content, with the kids doing cleanup, until one of my well-intentioned neighbors mentioned that hens are even happier with a rooster around.

I found a studly white rooster, easily twice the size of the largest hen and very aware of his superior status in the chicken world. He was just the thing to make the flock happy and had a menacing eye. I moved him in warily, mindful of his large, clawed feet and threatening curved beak. I was concerned about him overdoing the male dominance thing, but trusted his instinct. Welcome into rooster heaven I thought, as I thrust him into the henhouse.

The next day when we went out to see how our flock was getting on there was a most astonishing scene. Our stud rooster was cowering, disgraced in the corner, trying to hide from his harem! Every feather from his neck to his shoulder was gone and raw, with pink skin exposed instead of the proud mantle of feathers. Our rooster had been well and truly henpecked and lived in terror!

This rooster was in dire straits, and I confess certain male sympathy with his plight and irritation with the bullying and unwelcoming flock that had savaged their new friend. Our proud rooster was now an object of pity instead of a new leader of the clan. The rooster needed a pal, and I felt inclined to help.

After much family reflection we arranged a little experiment that made the difference. I fenced the henhouse into two sections, and allowed the rooster to recover his dignity and feathers alone for a few days. When all had settled down, I plucked out the smallest hen and brought her to the rooster side. The next day all was domestic bliss, with the hen and rooster best of new pals, and only a cluck, strut, peck, and croon on display.

A few days later, I introduced two more hens to the rooster side and the next morning again all was chicken bliss. A repeat a few days later, and the growing rooster side of the coop were happy campers. Finally the entire hen flock had been introduced and there was only a peaceful family clucking and strutting. The rooster even seemed to think he was in charge, with his neck plume now grown back and his pride restored. The fence was removed, and the flock lived happily ever after.

There are many lessons to be learned from this chicken story, but one outstanding thought occurred to Kate and me as we enjoyed our breakfast of fresh eggs. We were both very happy that there was only one of each of us.

· 🦃 ·

Absolute Worst

KATE AND I were sound asleep at our ranch when the phone rang at 1:45 a.m. The panicked voice spoke the worst words a rancher can hear: *"¡Señor! Fuego! Grande fuego!"*

I slammed the receiver down and threw the phone to Kate to call 9-1-1 as I jumped into jacket and rubber boots, the quickest ranch clothes possible. A minute later I arrived at the ultimate disaster. Our big barn had exploded into fire, with flames erupting through part of the metal roof. More flames leaped out of each stall door...*the horses!*

Jose Carlos and I ran to the barn. The heat blast from the stall doors turned us away. Flames poured out and we knew that nothing could survive inside. Three horses were doomed.

In minutes, the Fire Department arrived from Buellton. Later I learned that the first vehicle on the scene was actually the arson investigator. The firemen immediately and brilliantly moved into action, connecting hoses to the ranch hydrant and spraying water from their trucks, unfortunately with little effect.

There was no time, in the heat of the flames, to mount and try to start the two fuel-filled tractors, fortunately parked in the back of the barn and momentarily away from the approaching flames. We attached chains to the tractors' back tool bar, and in some danger, dragged them out. I well remember driving my four-wheel-drive Jeep, throwing a chain around the hitch, and pulling a tractor backward out of the open barn. Normally, firemen would never allow such private enterprise, but this was country, and the tractors were saved.

More fire engines arrived, but there was little to be done except contain the fire to the one big barn, let it burn, and hope to save the rest of the ranch buildings. Fortunately for the ranch, the containment was successful. The bulk of the barn was hay storage, and that part had exploded in flames. Otherwise, all that could be done was to gradually subdue the fire, with the burnt trash to be carried away the next day. The tack room section and stalls were almost all destroyed. As I write this account over twenty-five years after the event, I can still picture, smell, and even feel the heat and horror of the desperate few moments of trying to save what we could.

Our ranch was just off 101, and for an hour County 9-1-1 emergency dispatch operators heard over and over again from motorists reporting a barn fire as they drove by in the early hours.

We had hosted a large ride that weekend, and there were a number of horses staying on the ranch, although thankfully only three in stalls. Words cannot express the horror of losing a horse in a barn fire. The mutual bond between horse and owner, which grows to a unique and meaningful relationship beyond sentiment, and is the epitome of people-animal trust, had been totally violated by the brutal and lethal fire. Three horses were lost that night.

Traveler was a big bay gelding famous for making his owner, Bob Post, look like a great horseman. Bob was a senior partner with a heavy-duty Los Angeles law firm and probably not the most athletic man in the world, but very good company. Somehow Traveler and Bob won prizes in horse shows, three-day events, hunt events, and even steeplechases. They were a matched pair of gentlemen who took care of each other, the man with the horse trailer who paid the feed bills and took all the glory, and the horse, the honest athlete who never refused to jump a fence and kept his boss from falling off while he was jumping. Traveler was lost that night. Bob was not on the scene, but when he found out, he was devastated.

Dave Wendler was a professional horse trainer who often brought a string of horses for client riders to ride in our Valley events from the San Fernando Valley. He had great horses, but they were accustomed to living together in corrals, and his string was outside except for one particular horse that he thought was worth the extra care of a stall. The next day he sadly loaded his personal and clients' horses into a trailer for home, but he left the best one for me to bury when the ashes cooled. There were remains of the tragic animals in the stalls that, because of the streams of firefighting water, had been partially preserved. Except for one, and that was even worse.

Lee Swett is a great cowboy who then kept horses regularly on the ranch when he leased our property in those days to run his own cattle. I had given up my cattle ranching by that time for full time pursuit of the wine business. His great roping horse that he just called "Friend" was in the first stall, and was presumed to have perished with the others. Gradually as the fire diminished, we could see that the one stall door was open. We searched the fenced area of the house and barnyard and found an apparition.

The Sixth Chapter of the Book of Revelation, Verse 8, reads: "And I saw, and behold, a pale horse: and he that sat on him, his name was Death; and Hades followed with him." Probably no biblical scholars and few horsemen will ever behold or realize the actual horrors of a pale horse, but we did that fateful night. Lee had found his horse standing up against a back gate. He gently led the hurting animal with a loose rope back into the light to reveal a completely burned and hairless creature, with his skin a pale ashen color. The horse was in a complete state of shock, shivering and gasping for breath. In the last desperate moment, but too late, the horse had kicked open the stall door. Fortunately, Lee had a pistol, and we all turned away to allow him to do the only thing he could do to help his poor horse.

The next day we were all in a state of shock as friends and neighbors came to help us dismantle what was left of our barn. We buried the remains of the horses on the back of the ranch. There was nothing to salvage, as burnt tools and odd items were charred beyond repair. The fire investigators stayed all day to survey any possible evidence and closely inspected each bale of hay that was dragged free to search for spontaneous combustion. None of us smoked and there was no obvious cause, but of course there were many visitors the day before. Ultimately no evidence was found for the fire source. The mystery and also the horror of the fire remain with us to this day.

· 🎵 ·

Horse Hysterics

CHICO WAS an extremely handsome chestnut Quarter Horse who suffered from what a first-class horseman cowboy, Deacon Hobbs, described as "Too smart for a horse." He had started as a running Quarter Horse until he decided not to win, was a show western pleasure horse until boredom set in, and finally the over-qualified animal wound up on my ranch as a working stock horse who made me proud when he was in a good mood. When he wasn't in an easy frame of mind Chico was prone to fits of excitement and temper, and those moments could cause embarrassment.

One day, a bunch of us had gathered cattle into a corral about a mile from ranch headquarters, where we next separated some steers to be shipped out in a trailer. I was riding Chico but had dismounted to work a corral gate on foot while others cut out the desired steers on horseback. Our then fourteen-year-old daughter Hayley stood holding my horse outside the corrals with some friends who had ridden out in a Jeep to watch the action.

Two of our friends, Jim and Joanne Oelschlager, were visiting from Ohio and enjoying the western action. Jim is a financial wizard and a good friend, but they are, as you might describe them, city mice. It was a normal ranch scene, cowboys working the cattle in the corrals while horses and visitors watched, leaning on the fences. Then Joanne mentioned to Hayley that she had ridden some and asked my unsuspecting daughter if she could sit on Chico.

Anyone who has owned horses knows the problem of visitors wanting to ride. How do you establish if they know how to manage a horse if you have never seen them on horseback? On a guest ranch, there are certifiable safe horses that will only walk behind the horse in front, and those honest animals are prized for the work they do to take care of non-riders. Ranch horses are another matter, and most have their quirks that a greenhorn may not be able to handle. It usually does little good to ask. The person who has ridden a pony a few times in camp twenty years ago will say, "Yes, I have ridden all my life." The person who won the Pacific Coast Junior Stock Horse Championship will say, "Yes, I have ridden a little." You never know. Generous and friendly daughter Hayley did not see why Joanne could not sit on Chico and walk around a little while we were all occupied with the cattle sorting.

None of us in the corrals noticed as Joanne climbed on Chico and took the reins in her inexperienced hands. At first, the horse accepted a new rider but then, as she walked out, a misunderstanding developed something like the fate of nations leading into a world war. Joanne, to gain a firm seat, held onto the saddle horn and tightened her legs into Chico's sides. In the horsey mind, this meant go a little faster. Chico was bored and frustrated with standing around, and an excuse to move out was a welcome change. He trotted and Joanne clamped her legs tighter. When he broke into a canter headed for home, Joanne held on even tighter, and then all hell broke loose as those of us in the corral finally noticed the runaway.

Home ranch corrals were about a mile away, up a dirt road, most of which was cut from a hillside with a steep drop-off. A gooseneck truck and trailer were driving down this road to pick up the steers. Chico, with a desperately holding-on Joanne, ran up the road past the rig at a flat-out quarter-horse race gallop while truck driver, cowboys, Hayley, and Joanne's husband Jim gaped. Chico's mind had one thought: "Run home fast!" Joanne was in desperate trouble. Some of us jumped on horses to follow, but slowly at a distance, because there was nothing we could do to get in front of the train wreck up ahead.

Now came the miracle. A cattle guard lay at the end of the road to the home ranch, which is a pit dug in the road about six feet broad with iron poles laid across so vehicles can pass, but horses and cattle cannot negotiate the gaps between the poles. The road was cut into the hillside with a bank on one side and a drop-off on the other, and there was no way around the hazard. Chico headed for this obstacle a hundred miles an hour, with the hysterical Joanne still holding on. Not fifty feet before the cattle guard Chico must have side-stepped because Joanne was thrown off and rolled down the bank, which fortunately was loose dirt. Miraculously the steep angle and soft soil had cushioned her fall, which otherwise could have been lethal. She came to rest against a tree growing out of the hillside and, against all odds, emerged battered and bruised but not broken or worse. Chico was not a jumping horse, but somehow had managed to fly over the cattle guard with one leap and ran to the stable where he stood shaking in a muck sweat.

We arrived shortly after and ran down the bank to find Joanne covered with dirt, bruised, slightly bloody, but smiling and without major injuries. Like all good guys, her emotions were more embarrassment than grief or panic as we helped her up the bank and attended to her minor injuries. The grief and panic were all ours.

Sometimes when I wake up in the small hours and the night demons send me dark thoughts, I still recall this hysterical ride as the worst horse moment of our ranch life and my dreams are invaded by what might have happened.

· 🎠 ·

Ghost Rider

SOME TRAINERS believe that nobody will ever teach a horse anything because it is not their nature to think. They maintain that horses only acquire habits and do what they do for better or worse by repetition. However every once in a while something will occur to shake up this line of reasoning.

Don Hansen ran the old Jack Mitchell ranch across the river from Santa Ynez town. He needed a working ranch horse, and was lucky to buy a very superior animal named Winkle from a first-class cowboy who died shortly after the deal was made. The horse was gentle but had won shows, cuttings, and ropings in his day, and was now too old for competitions. Winkle proved to be very adept at accomplishing normal ranch chores and was a pleasure to be around. Don kept Winkle in a paddock with another cow horse usually ridden by his wife Sue, and the two animals had become good friends and accustomed to each other's company.

One day Sue was away, and Don saddled up her horse instead of Winkle to give him some exercise and accomplish a simple job of bringing in a few steers that were in the wrong pasture. Of course there is never a given success about moving steers and many things can go wrong, but Don figured it would be an easy chore and a routine job on the ranch.

As Don rode out, Winkle tagged along and somehow got out through the gate and into the open before Don could close him in the paddock. No problem here and he decided to let the horse tag along. Roping and leading the loose horse or otherwise persuading Winkle back into the pen was more trouble than it was worth, so Don figured no harm could come from letting a horse follow along on the ranch. If he wandered off there were fences, and he could not get lost on the spread. As it turned out, the loose horse just moved with his stable mate and the three, Don riding Sue's horse and Winkle, just walked along together.

Soon they came up on the few steers that needed to be brought in, but the steers apparently were of an independent mind, and the herding did not seem to be coming along. Don was having trouble bunching them up in the right direction, much less moving them back to pasture. With that, the loose Winkle split off from Don and charted his own course around the steers. The exceptional horse just seemed to know what to do, and together, with horse and rider on one side of the steers and riderless horse on the other, the cattle herded toward the gate previously opened to bring them into the right pasture. They moved slowly at a walk, but occasionally Winkle would trot back and forth to keep the herd together and he moved just as he should do and as if guided by an experienced cowhand. The loose horse made all the right moves and the job of herding the small bunch was soon accomplished.

Don accepted the help and appreciated that the job was done, but was astonished at having a lone horse as a partner. When the amazed Don closed the gate on the steers the three, two horses and one rider, walked back home, unsaddled, and soon Winkle and his friend were munching hay in their home pen and Don was scratching his head.

As was his habit, Don eased into a porch chair and lit an habitual cigar. He contemplated the extraordinary ranch scene he had just witnessed and tried to sort out the meaning of it and what to tell Sue. He accepted that the job had been done, but even years later wondered if the horse had really accomplished his half of the herding or if there had been a ghost rider, and maybe his former boss, telling him what to do.

· 𝒥 ·

Cow Bliss

FALL SEASON in the Valley is always a scarce cattle feed time with the dry, gold grass mostly eaten down. Ranchers who keep their cattle over the summer and into the winter hope to have saved enough feed on the hills and pastures to last until the early winter rain brings fresh grass, and until then, cattle are scrounging for food. But this time of year also coincides with a cattle-feed bounty from the wine-making cycle, and, as we had both a winery and a cattle operation, we were grateful to have a feast from the vintage for our herd.

Harvested grapes are dumped into the wine press that squeezes off the beautiful juice, leaving behind a mass of skins, seeds, stems, and the occasional leaf. This excess tonnage then moves from the press by conveyor belt out to an awaiting dump truck for disposal. In our case, disposal meant trucking the load out to pastures of eagerly waiting cattle. The load would be fresh, grape-sweet tasty, and somewhat nourishing, and also a magnificent change from dry grass. This dump trucking duty was a coveted and rewarding winery job accomplished by the boss whenever possible.

Inside the winery during harvest the work is hard, loud, continuous, and arduous. We always seemed to be physically tired with the long hours and many tasks. The relief brought from the grape disposal duty simply involved a quiet drive to the pasture, usually toward the end of the day, raising the back end of the dump truck and slowly driving along, depositing a steady stream of slushy, crushed grapes. Thinking of those quiet moments in the pasture brings back very happy memories. It was pure pleasure to leave behind the winery work and noise to enter a peaceful, quiet field and feed a herd of greedy cattle.

The frolicking animals would run to the truck with their hopping, abandoned stride, push and shove each other, and then chomp to their hearts' content. Cattle were also a relief from wine connoisseur critics and difficult customers because cows always approved of our winemaking and had no preference for Chardonnay or Cabernet, enjoying each vintage equally. No steer ever sent back a load of crushed grapes, unlike the picky customers who sometimes rejected a bottle of wine! Once we sent a sample of the crushed grapes to a cattle nutritionist, whose analysis confirmed the slush was high in sugar content but low in protein, with some good nutrients in the seeds. The cattle just enjoyed the sweet swill.

And for even more cow enjoyment, there were the fermented grape leftovers. The red wine process includes the juice going through fermentation along with the skins and seeds and then being pressed. The load after pressing would be a similar mix to the unfermented solids, but this time containing the full alcohol content of wine, with a pungent vintage aroma emanating from the dumped slush. Cattle enjoyed devouring this load even more, with a different result than the high-energy sugar content of the fresh load. After consum-

ing the load, a happy herd would then contentedly and sleep-
ily find a shady spot and digest the fermented meal in quiet
meditation. The truck driver would return to the hustle of the
winery, but the cows would settle in true cow bliss.

· 𝒥 ·

Horse and Glamour

A FEW YEARS ago, I needed to sweep our largest holding
pasture, which included about a thousand acres of hills
and valleys containing two hundred steers to be moved. As
usual, I asked around for help.

I reached our pal Chuck King, a top hand, hoping he would
join in, and he agreed, and asked if he could bring a woman
friend. "Does she know what she is doing?" I asked. "Well,
pretty much," he replied, "and she sure is gorgeous!" I asked
her name. "Bo Derek," he replied. Without hesitation I said,
"Bring her!"

On the appointed day, trailers arrived before dawn. The
idea was to ride to the back of the ranch by first light, then
gather in the pasture in the morning chill before the sun had
convinced the cattle to brush up in the shade of oak trees. A
thousand acres is not that big for a roundup, but there were
two steep canyons and a number of tricky side hills to hide
cattle. We had about seven riders from our family and ranch
hands, and another eight had planned to help. Of course I
would return the favor to those who needed similar help.

My first morning chore was to get the family out of bed,
and my older son and daughters presented the usual challenge. However the then ten-year-old Andrew, inspired by the
thought of riding with a beautiful star, bounded out of bed and
headed for the corrals with shirt tucked in and hat straight, a
picture of cowboy style. The rest of the family staggered down
to the barn to saddle in stable lights.

I noticed that Bo and Chuck had arrived and seemed to
be organized. When we were all ready and congregated in the
corral lights for a moment of coffee, Danish, and gossip, we

could see our breath in the dawn cold. The bundled-up riders set off on the dark road with hands in pockets.

We reached the back fence at sunup and designated the sectors to be ridden to various riders. We agreed on timing and strategy to bring in the herd across the rough terrain. We would separate and not see each other again for perhaps an hour or more until we began to view others bringing in cattle they had found and similarly heading them toward the appointed gate at the headquarters corrals. In the division of riders, I just had to have Bo riding in my sector with me, and of course Andrew joined us

Bo was well mounted and had, as we say, a cow sense to do the job. Additionally, she was startlingly blue-eyed, rosy-cheeked beautiful, and pleasant. The work began, but the country morning seemed to fly by too swiftly. The cattle became almost incidental to the joys of gathering a pasture with such delightful company. Kate had taken another sector with all amazing wifely sportsmanship, trusting Andrew to bring his Dad and Bo in safely.

As the sun rose and brought welcome Valley warmth, we soon found some steers and headed them in the right direction. Eventually other riders pushing their cattle formed up out of the hills and canyons. The morning went without mishap, ending up with the herd bunched together in the proper corner of the pasture ready to go into the corrals, a sight to gladden any rancher. The count, when the herd passed through the gate, came out right, with no missing cattle. The steers were cut out and separated in the corrals and went their designated ways without incident, and all the ranch tasks accomplished.

Bo may be a star, but she handled the roundup with a modest competence and unassuming help to the crew that won our respect. Later, I would see her in the film *Bolero* riding a show-off Andalusian stallion that would challenge any cowboy, and I realized that she was a first-class rider, but that day she just did her job. We accomplished our roundup mission and settled down to the barbecue lunch that was held out as bait for the volunteers. Bo was as pleasant company then as she was help on the range with, as they say, a smile that wouldn't quit. It was very clear why she is such a popular Valley fixture.

Later, when our guests had left and while we were washing off the horses, Kate reminded me that the morning had only been the stuff that dreams are made of!

Our ranch was graced with a beautiful rider again when Jane Fonda helped us gather cattle. She had accepted my invitation to help, ironically, when Kate was out of town. So I had Jane all to myself, except for her then-husband, Tom Hayden, and their son, Troy. My two teenage daughters were also riding out with us, and they made it clear that I was not to bring up any political issues. This was probably a good thing because later Tom and I served in the State Legislature on opposite sides of the aisle, and it is also doubtful Jane and I would have found mutual enthusiasm for world issues.

Jane rode Kate's horse with style and competence, and Tom rode a safe horse with my daughters. Of course Jane and I had to ride a distant sector together, and we had a great conversation bringing in our cattle. Our son, Andrew, and Troy rode with a ranch hand in the Jeep. The Hayden family were good enough riders and good company, but were probably more suited to show business and political pursuits than to ranching. It was a beautiful day, and, once again, Valley animals served to bring us all together in purposeful and amicable harmony.

· *F* ·

Dodge Solvang

CATTLE RANCHING requires moving herds of cattle from one place to another. Sometimes the feed in one pasture is used up and the herd needs to be moved to better fields, sometimes cows and calves need to be separated, sometimes the bulls need to be moved to the cows. There is always something to be moved around on a ranch. In the 1800s, the great cattle drives moved large herds, with great danger and difficulty, across states and plains to market railheads in dusty towns with romantic names like Dodge City or Abilene and over routes like the Chisholm Trail. After months on the trail, cowboys needed to remove layers of dust with questionable whiskey, lose their wages at the poker table, and sometimes shoot at each other to express the enthusiasms of time off. This is the stuff of legends.

The legend lives with regular cattle drives around the Valley, mostly out of sight of the growing population. Driving cattle is a skill, often learned through herding disasters when the cattle scatter or miss the desired fence opening and head back for the hills. Good herders have a cow sense and are able to know what the cattle are intending and whether a cow is thinking "Okay, show me the way and I will just mosey on" or "If you give me an opening, I am out of here!" Ranching families and ranch hands get to know each other's skills and also know the abilities of neighbors who come to help.

Anyone who wants to learn the joys of herding can go to a rodeo and watch the team penning event. Here, three riders race the clock to put three designated steers in a pen. The contestants with cow sense are clearly evident, as are the greenhorns who leave the ring shaking their heads in shame and frustration.

In 1974, we had about a hundred heifers that needed to be moved to a leased ranch in Buellton at the mouth of Ballard Canyon. This was a question of hiring a truck, making many trips in a gooseneck trailer, or simply herding the animals. With the advice of a couple of skilled if overly optimistic cowboys, I chose the least wise plan, and with cowboys, family, and a couple of neighbors, determined to make the cattle drive. A reconnaissance of Ballard Road indicated reasonable fence containing the road, and at that time there was little traffic. Years later, I still can't believe we attempted this.

We saddled up and let the herd out to set off from our first ranch above Los Olivos and drove down Foxen Canyon to cross Highway 154. We traveled without benefit of Sheriff or Highway Patrol, who would probably have put me in jail had they known what we were up to. We simply put out mounted road guards and stopped bemused traffic until the herd had crossed 154. Getting up the hill above Los Olivos to the top of Ballard Canyon was a trick because the heifers did not like the climb and we had to do a little yelling and urging. We put mounted family in all the driveway openings and managed to keep the heifers on the road.

They scampered down Ballard Canyon, where the Amgen Bicycle Race now takes pride of place, held to the road by fencing along the way. Our challenge was to parcel out enough riders ahead to wave off traffic and keep the herd bunched up and steadily moving in the right direction. The Valley was a little more country in those days, and the few drivers we passed were content to pull over and watch the show. The cars held up in back of the meandering herd seemed to enjoy the fun.

We had a little bit of a circus act at the Chalk Hill fork when they milled at the side of the road and seemed to want to return to Los Olivos, but finally we made the gate and the herd set in happily on the lush grass of the new pasture. We rode back home pleased that we had taken a chance and got away with it and thinking we would be able to tell our grandchildren about our Ballard Canyon cattle drive, an event that would not likely be repeated. We now fully realized that street drives could be dramatic, such as the famous Solvang cattle drive.

In 2008, the Rancheros Visitadores decided to add a cattle herd to their annual parade through town to the Santa Ynez Mission. This was not as adventurous as it seems, because drives through towns are often part of the opening of rodeos, of course usually in communities in Montana or Oklahoma that are little more attuned to the ways of the West than downtown Solvang. Ranchero membership contains some of the best cattle people in the business, and the best of these top hands got together to plan the parade drive.

Cotton Rosser is a premier rodeo manager and stock supplier to the shows. He kept a mature herd of Longhorns with a good attitude that had paraded down main streets before and knew the routine. Gil Aguirre is one of the foremost California ranchers, with a long and successful history of competitive roping and horse events. When these two formidable men got together, they selected about twenty more from the

Rancheros group to herd the cattle through town. Together, this group of primo cowboys probably had more championships, trophies, and silver buckles than any other group twice their size.

On the big day of the annual Ranchero parade through town, they formed up the herd in a set of corrals below the town and prepared for the drive. There were two situations working against them. Firstly someone had decided to add some local steers to the herd to fill out the show. Normally, this would be no problem, because the new cattle would likely just settle in with the big, mature Longhorns. The next problem was the afternoon breeze which sometimes tends to make horses and cattle a little goofy. Western lore is that wind spooks animals because they sense it is carrying their scent to the predators and their mind-set is to run away from trouble.

The parade started off from the canyon below the Mission with the old-hand Longhorns, younger somewhat confused steers, and posse of champion cowboys. The herd made it across Alamo Pintado Creek in good order, and up the steep hill behind the old Scandinavian, now Quorke, Hotel. When the herd came to make the right turn up Alisal Road and into the main section of town, confusion set in. The herd lost cohesion and direction, and the cattle, abetted by the wind, wanted to go somewhere but did not know where. The cowboys watched in amazement as their herd became loose individuals and sought refuge in driveways, gardens, and front lawns. Homeowners and tourists watched in amazement as the scene became, as the cowboys say, "A Western moment."

The herders applied all their individual skills to repair the confusion. There were shouts of "Howdy, Ma'am" and "I'll have him out of your roses in a moment" and "What is that animal with the horns doing in my garden!" and, most of all, "Sorry for all this and I hope we didn't trample anything!" The regroup probably lasted five minutes, but to the red-faced cowboys it seemed like an hour and went down in the Solvang history book.

Finally the herd was back together and off in their glory to the waiting crowd and reviewing stand. But the champion herders had one more red-face moment. When after they started off again down Alisal Road, three additional steers emerged from someone's driveway and followed the herd to catch up. Apparently some kid from the crowd called out, "Mister Cowboy, wait for your cows!"

At the Mission, the good Father Michael and dignitaries were waiting to give the speeches and blessing, but Gil Aguirre thought they had best move on. "Don't you want the traditional blessing?" the parade chairman called out. Cotton Rosser called back to the awaiting priest, "Padre, if you don't mind, we had better get this gang out of Dodge!" And the herd departed through the Mission parking lot and down the hill into history, as this might not be tried again.

· 🎵 ·

Decoy Duty

STEVE AND Katherine Pepe were important lawyers in downtown Los Angeles in a former life before deciding to move to the Valley and the vicissitudes of making great wines for the rest of us to enjoy. Clos Pepe has a first-class wine reputation and the Pepe family is happy in their work, living their dream in the vineyard on Highway 246, halfway to Lompoc. Perhaps even more pleased with the move to the country was the family Dalmatian, Bud, who was totally content with his free life on the property and away from city constraints.

However, Bud found himself engaged in a country ritual on the new ranch that played out over the years with a recurring plot of dog frustrations. In the quiet of a sunny Valley day, Bud would lie in the backyard gazing at the large reservoir pond that provided irrigation and frost protection water for the vineyard. Dalmatians are good swimmers, but a plunge in the water did not interest Bud nearly as much as a duck dinner.

For many months of the year, the pond would be home to a flock of Mallard ducks that traveled in from the cold north and appreciated the vineyard pond as much as the Pepe family. There were rushes and calm water and every accommodation

for visiting ducks to enjoy. The bright green males swanned around in the water and the gray-brown females were happy and at home. The duck families were safe and content as long as they kept an eye on the Dalmatian.

Bud would balefully watch the ducks in their bucolic setting and then periodically his thoughts would turn to hunting

134 • *Valley Animals*

and a duck feed, and his sleepy ways would suddenly turn to an aggressive search-and-destroy mode. Without warning he would charge down the bank and plunge into the water, swimming flat-out for the duck convoy. At this, the ducks would circle the flock with the females inside and the males forming an outer, protective ring. When the thrashing Bud began to approach the flock, one of the males, chosen by whatever duck selection, would swim off sideways, beating the water in a distressful manner away from his fellow Mallards. Invariably, the Dalmatian would pursue the thrashing lure, away from the watching crowd.

The single decoy duck had only to keep a wary eye on Bud and paddle well ahead of the dog teeth, which was easily done. A variation to the fruitless chase might be when the dog was led under the inlet pipe where the well water was sprayed into the pond and had the effect of disorienting Bud, sometimes causing him to give up the chase altogether. If for some reason Bud came closer than duck safety allowed, the green Mallard would simply take off flying and disrespect the dog from the air. If the duck otherwise tired of the game, the single thrasher would return to the flock and another male would take over the decoy responsibility. Inevitably, and time after time, Bud would tire of swimming and forget about the duck dinner to retreat panting to the backyard and doze the incident off.

This same scenario played out for over ten years with the chase tirelessly repeating itself, albeit with Bud's pursuit slowing down some in his later years. New visiting Mallards would employ the same winning tactics and Bud would experience the same frustrations without ever giving up his vision of catching a duck. Finally Bud passed into the dog beyond, and the Pepe family is sure that the Mallards still look for the dog on their visits and miss the sport.

· 𝒥 ·

Part Five
Domestic Partners

Shaggy Matchmaking

VALLEY FOLK, and particularly those who dwell on rural byways, know too well that dogs and cats are often left abandoned on the roads. These deserted and lonely creatures sometimes find a local home and sometimes are forced to move on. As our family has lived on well-traveled roads, we have acquired some great dog and cat friends this way. Of course these orphans need to have the right disposition, be looking for a family, and be able to make an accommodation with the animal occupants in residence at the time.

The Dunn School, located on Highway 154, has experienced a number of wandering animals attracted to the wafting aromas from the kitchen and a young population willing to give affection to strays. One day, a generation ago, a friendly Golden Retriever, starving and emaciated, wandered into the embrace of the Dunn students. An abundance of junk food and the administration of much spoiling and petting brought the dog back to health and spirits in short order. His golden coat glowed, his eyes brightened, and his tail wagged. He was a loving and energetic campus character, but he soon ran into difficulties with a school policy that threatened his good life. Students were not allowed pets on campus and his ownerless status exposed the poor Retriever to expulsion.

Dunn management advertised and posted notices, but nobody claimed the stray dog. And then his threatened exit was saved by a technicality. Faculty living on campus were allowed pets. A popular bachelor on the staff, Doug Jessup, who had grown up with family Retrievers and took an immediate liking to the orphan, claimed the animal as his and they moved in together. Of course the stray, now named Jason, continued to wallow in the affection of the entire campus before retreating to the bachelor apartment to curl up in legal residence on Doug's floor.

But Jason had been abandoned once, and who knew what doggy dread lurked behind the friendly canine face when Jason might sense the possibility of becoming alone once again. Could Jason have somehow felt the need for a more permanent family than only a single teacher? Do Golden Retrievers long for extended family stability with an instinct as sure as the need to flush birds from the bushes?

Meanwhile, another attractive and single teacher named Hayley Firestone was also part of the Dunn school faculty and had an affinity for Golden Retrievers. In the course of their professional lives, the two teachers would occasionally meet, and always present was this flaxen, wooly friend, who would hover and flop and emit loving dog vibrations. Hayley would always be welcomed with the thump of a tail on the floor and a warm dog look. There is more to heaven and earth than we know, and who can deny, define, or discount the ability of a loving dog, yearning for a home, to promote magic in the relationship of two compatible, eligible people—people very capable of providing a permanent dog home.

Doug's office was in a woody, comfortable space because he was teacher, coach, and college counselor and had frequent meetings with the students. He and Hayley also began to find time for meetings while sleepy-eyed Jason looked on approvingly. Soon the meetings were facilitated with an inspirational glass of local wine and the glow of a wood stove while the furry animal lolled in a comfortable mood of familial contentment, willing the future in his homey, doggy way. Distant memories of homeless life on the open road surely moved the loving Jason to promote a permanent establishment with these two in every way possible.

In good time, the natural course of events proceeded and the conferences became more lengthy, personal and harmonious despite the curious eyes and gossip of the school. In due course there was a spectacular wedding, a home every dog might wish for, and then later, three sisters and a brother to play with as the family grew. A shaggy dog's matchmaking dreams were realized to perfection.

· 𝓕 ·

High-Tech Rescue

WHILE SWIMMING in the home pond, the three Walker girls, Ella, 14; Georgina, 13; and Tamsyn, 12, were aggressively dive-bombed by a small, brightly colored gray-and-orange-feathered missile. The small bird was clearly disoriented and behaving very strangely, perhaps thirsty for pond water, or upset by a bird reflection in the water, or wanting and then rejecting the company of the swimming girls. Parents Polly and David soon got into the act with sympathy and good intentions for the creature, soon identified as a very upset and distressed Cockatiel.

The family tried to help and rescue the bird from its frustrations with hands held out, bread crumbs, and sympathetic noises. For over half an hour, the Cockatiel teased the family, first pretending to land and then zooming off to the branches of an oak tree. Even there, he would set out toward David, who was attempting rescue on a ladder, and then waddle back or fly off. Sometimes he almost landed on an outstretched hand but then seemed to reject the person as an unacceptable stranger.

Finally Georgina had a brainstorm. Using her laptop she Googled "Cockatiels" and found a two-minute YouTube recording of chatty Cockatiel birds tweeting with each other on the computer. She turned the volume up and played the recording continuously. The flirting and curious visitor was drawn to the sound, circled, landed, and waddled closer to the bird sound on the computer. Sister Ella was ready, and a gentle towel thrown over the beautiful little bird decided the issue. Delicate hands transferred the Cockatiel to an empty box, and the bird was now domestic again.

Meanwhile, sister Tamsyn had phoned a good friend whose parents proudly and affectionately kept two Cockatiels of the same coloration in a cages, happily side by side in their home. Tamsyn believed her friend's family, Mr. & Mrs. Roots, would surely know how to proceed in finding the rightful home of their visiting stranger. And then by good fortune a home was secured.

Tragedy had taken one of the Roots' pair of birds, and the survivor had gone silent without a friend to twitter with in a now neighboring empty cage. The visiting stranger was brought from the Walkers and set in the empty space of the departed friend, and the two immediately set off singing in bird harmony. The two families dutifully published appropriate lost-and-found notices and spoke with pet stores, but the original owner of the stranger bird was never discovered. In the Roots' home, the two Cockatiels live together in adjoining cages, making conversational music thanks to the computer savvy and bird sense of the Walker girls, who were very chuffed to find a home for the distraught dive bomber.

A younger generation brings to effect a new high-tech relationship between Valley animals and people that old-timers might have trouble understanding.

· 𝓕 ·

Home Sweet?

TUCKER, THE shepherd dog, Kate, and I live happily in our home with Cricket the cat. Cricket is an alumnus of Catalyst for Cats and she decided years ago that we are her family. The Catalyst people rescue stray cats, bring them to health, neuter them to prevent more strays, and find them good homes.

Typically, a Valley household will notice that a semi-wild, feral cat is multiplying with litters of kittens, threatening overpopulation, and call Catalyst. If the cat is tame, a simple trip to Catalyst will spay or neuter the domestic cats. A humane trap will serve to bring in the wilder feral cats for the same easy treatment. Unless a household specifically wants to breed and reproduce pet friends, they should all be spayed or neutered to prevent the hardships of overpopulation and all the strife and suffering that results from too many animals on the street. Catalyst for Cats will bring in around seven hundred cats a year from the County, including something over a hundred from the Valley, and return the cats to healthy, productive lives. They are one of the stellar groups helping Valley animals, and a tipped right ear for the males and a left ear for the females is an indication of Catalyst alumni.

One day, looking for a resident cat friend, we called Catalyst and they were kind enough to deliver a gray-striped female to our ranch. We called her Cricket, and she lived in a cage for her three weeks with us, to become familiar with our premises. When Cricket was let out, she adopted our home, and the pesky mice that occasionally minced their way across our kitchen vanished and moved to other ranches.

I have never known a cat to talk politely to me, but Cricket will say hello when she wants attention, say something more definite when it is mealtime, and sometimes just meow to be friendly. She loves conversation and back-scratching. Shepherd dog Tucker is equally surprised to be good friends with a cat. She will sleep with Tucker and sometimes flex her claws in his mane. This must give him some sensual pleasure, because he adopts a dreamy look under this treatment, with a vague expression of nothing going on. Kate assumes that cats will like her, but Tucker and I are amazed that Cricket seems to like us too.

Of course cats have dimensions of violence and mayhem that are beyond the understanding of less aggressive dogs and dads.

One quiet afternoon I was engaged in my favorite occupation in my preferred spot. Kate and I sing with the Santa Barbara Choral Society and work hard to keep up with this accomplished group. At that time, I found myself happily perched before a picture window, mountains in view, with a cup of tea, practicing Handel's Messiah by singing along with an instructional CD in preparation for our upcoming Granada performance. I was at peace with the world and quite settled.

142 • *Valley Animals*

Suddenly just outside the window along came Cricket with a struggling rat in her mouth. It was a large wood rat with a large wood rat's long tail, and it was very distressed. Just before my eyes, the two fell into wrestling, and the rat escaped. The chase led under some bushes and behind some outside chairs. Cricket emerged with rat in mouth and got down to the business of killing it just in front of the window. So far the struggle had taken some moments and caused me to miss a few measures of Messiah practice. Music was in suspension, tea cooling, and a sense of unreality was creeping in.

Fortunately the lethal moment passed quickly, but the cat then teased the new cadaver in a useless game of trying to continue the struggle. Now I had to decide whether to sing or retreat. I sang.

Here I was, a peaceful man singing a heavenly composition of great beauty, and the cat now started to consume the erstwhile rat just outside my window. Yin and yang, as they say, weren't half in it, but I did not retreat. Perhaps my experiences with the realities of a working ranch and all the attendant moments of life and death in the course of cattle ranching had hardened me to this domestic violence. After all, animal ways are as much a part of our world as great music. The rehearsal continued knowing that Cricket was on the job and a good friend to the family homestead doing her thing.

· *𝒥* ·

Cold Pizza

ANYONE WHO has been closely acquainted with a Jack Russell Terrier knows that there is personality, energy, single-mindedness, affection, and loyalty all rolled into a small dog with a very big-dog outlook. Jack Russells are almost always good looking, with qualities of a lap dog but also the ability to snarl, reminiscent of a junkyard guard dog on the fight.

Years ago, daughter Hayley and I secured a Jack Russell

puppy fur ball from a friend in Pebble Beach as a gift to my wife Kate, the dog to be known thenceforth as Pebble. The little mite contained all the attributes of Jack Russells in heroic dimensions, and was very much part of the family for many years. The local vets sewed her up when she got into scraps, although she usually gave worse than she got. But her favorite spot was Kate's lap.

One day our extended family and a few friends decided to have a grand pizza party at the back of the ranch. Son Adam's date was also named Kate, who our whole family thought was a better girlfriend than he deserved. They had become very close (in fact, now they are married with four children); and at the time we were all rooting and hoping for the couple. As was typical of girlfriend Kate, she volunteered to pick up the pizza we had ordered while the rest of us drove to our favorite ranch picnic spot, which was beyond the reach of any cell phones. We all drove off to party and Kate drove the pizza run in our family car, hardly noticing that Pebble, not to be left out of the action, had ferreted her way into the vehicle. The two drove to town happy as you please, although they did not know each other very well.

In the Solvang Pizza House, Kate picked up the five extra-large pizzas and returned to the car and to a very intimidating, vicious, and snarling Jack Russell. The windows and doors of the car were closed, and Pebble had taken over and taken charge. She was not about to let this person who was not part of her family anywhere near the car, even with pizzas. Her large jaws and raving, killer roars were enough to stop anyone at the door handle. Her hot breath fogged the windows and that car was hers.

Adam's girlfriend was now in the pizza car park, new to the Valley, and with the assembled folk wondering why she was trying to break into the vehicle this brave dog was protecting. She phoned Adam and the rest of us, but we were well out of cell contact on the back of the ranch. None of the parking lot onlookers had a suggestion or wanted any part of the monster inside the car. The pizzas rested on the hood and Kate, now in a serious predicament, wondered what next?

Finally, after a number of false starts, she reached our daughter Hayley on long-distance. Hayley was able to contact our good friend Nancy Carter in Solvang, a person wise to the ways of Valley animals. Nancy had known Pebble for years and when she quickly arrived on the pizza parlor scene, was able to look the guard dog straight in her Jack Russell eyes. "In the back!" Nancy roared, louder than Pebble, and these were words and a tone Pebble understood. Some people have the experience, ability, and confidence to do that, and it worked immediately. Pebble retreated to the back seat, Nancy got in the passenger seat, and a very shaken Kate drove the semi-warm pizzas to our party.

Thus Pebble was able to attend the party, Nancy joined the gathering, and girlfriend Kate retold the amazing guard dog story and never trusted Pebble again.

· 𝒥 ·

Mother Love

A MATURE 14.2-HAND roan Quarter Horse named Pepper came into the Walker family for the eighth birthday of daughter Ella, who was a good rider and crazy about horses. The two would regularly attend Pony Club Saturday lessons and competitions and Pepper became a much-loved member of the Walker household.

The Santa Ynez Valley Pony Club fits the rural nature of the Valley and has been a fixture for generations, teaching young people about horses, sportsmanship, and life, in the best National Pony Club tradition. Many Valley young have grown up on the regular meetings, mostly due to the kindness of the Shepherd family, who offer their ranch just off Highway 154 for the activity. Pony Club is to riding as Pop Warner is to football and AYSO is to soccer.

Thanks to the Pony Club, Pepper and Ella did well on the young equitation and jumping circuit. Pepper proved to be a very talented small horse, although she had her ways and the rider needed to exercise constant attention. Ella trained and groomed and grew up with Pepper as a challenge and constant companion. Her two younger sisters looked forward to the day Ella would outgrow the small mare, and they could take on her winning ways, but that was not to be.

One morning Ella's mother, Polly, noticed Pepper taking a strange stance in the paddock and then staggering toward the feed. Ella had mentioned that the horse had not been quite herself in previous days, and this morning there was a real problem. The Walker family were barely able to load her into the trailer, and the vets seemed stumped to diagnose her instability and uneven walk. Pepper's eyes lacked focus and even with some sedation she was nervous and fretful. The Walker family experienced the horror of seeing a good and trusted horse friend stagger about in threatening distress.

Finally, after examinations and consultations, the conclusion was reached that the stylo-hyiod bone in her jaw had somehow been injured, with the resulting infection and inoperable fracture permanently undermining her stability. The Walker family faced the realization that she could not be ridden again.

Pepper returned to Ella from her stay at the vet's, improved but still not a whole horse, and, of course, unable to do the Pony Club rounds. Gradually she recovered enough to be bred to a very likely Welsh Cob, to accomplish what she was still able to do. After a brief honeymoon, she returned to the Walker's, happily in foal, but tragedy struck again. Her time had come to give birth, but instead, she was found one morning by the girls with a stillborn foal in her stall. There was no explanation and it was just one of those horse tragedies. Now, Pepper was very ready to be a mother, with her broody disposition and full of mother's milk, but there was no foal and she was clearly unhappy.

Polly Walker began phoning every vet and breeding farm in the Valley to see if there were any orphans, and at last there was a turn of good fortune in the sad horse story. The Travelers Farm is a very serious cutting horse breeding and training operation in the Valley. A champion cutting horse can have an importance and price equal to a race horse, and Travelers breed, train, and show the best. One of their champion mares, bred to a champion stud, had suddenly come down with terminal colic and left a potential champion foal an orphan. They had tried every formula of sweet milk administered with bottle and nipple and other nourishing mash treats to tempt the newborn foal, but nothing seemed to be working and the youngster was going downhill. When they spoke to Polly and learned of a mare looking for a foal, the Travelers' deluxe trailer was at the Walker door in minutes.

At the cutting horse ranch, Pepper and the orphan newborn filly at first eyed each other with mutual suspicion in the stall. They circled and sniffed with confusion and without connection. The Travelers' staff looked on with trepidation, hoping for the best for their young champ, and urged Pepper to make the bond, nudging the two together. Finally a hungry foal looking for comfort and a lonely mare with a bursting udder followed nature's way, and the two joined forces. The foal was happy with adoption in the new world and Pepper was totally content to offer horse mother love. From there on, the

two were as close as any mare and foal, and the young filly prospered in Pepper's tender care. Soon they joined the other pairs of mares and newborns in the Travelers' green pasture in total contentment.

Eventually the young filly grew, was weaned, and went on to cutting horse glory, and Pepper returned home with a horsey sense of the circle of life accomplished. The story has an even happier ending. Another trip to the stud horse was arranged, and Pepper produced a brilliant filly of her very own to be named June Bug, who is now the joy of the Walker family and winning ribbons and trophies for Ella's sisters. Pepper is retired these days and a happy camper horse, if still slightly off balance.

· 𝓕 ·

Special Relationships

TOM LEPLEY swears this story is true. It probably is, and it only illustrates the unfathomable dimensions of Valley animal minds, especially cat minds.

When Tom and Maggie Lepley moved into their home on Refugio Road with their four daughters, then age two through twelve, they acquired a couple of cats in the deal, Bowser and Miranda. In Tom's eyes, the cats were ugly and indifferent and the cats seemed to think the same about Tom. The four girls got on famously with the resident felines. The girls loved the cats and the cats knew this and put the relationship to good use. The cats put up with Tom but knew they could get good feed and affection from his daughters, and the mixed relationship continued.

The cats had an unfortunate habit of clawing the outside furniture. One day Tom caught Bowser savaging a chair seat and threw him over the patio onto the lawn where he landed on his feet, as cats do, and sulked off to despise Tom even more. But mostly, the cats seemed to ignore Tom as a supernumerary in an otherwise cat-loving household. The cats did have the advantage of keeping country mice away and they seemed harmless enough, luxuriating in the Lepley home, and so the months passed, with Tom and cats just tolerating each other.

One day Tom returned from feeding his horses to notice Bowser sitting and staring, absolutely without motion. There was a threatened, electric feeling about the cat. Tom stopped to observe, and soon noticed a rattlesnake coiling in front of Bowser and beginning the eerie dry buzz of the deadly rattle. Cat and snake were otherwise motionless and dread was in the air. Tom watched, also without motion. His daughter, Anna, appeared at the door and also took in the scene with horror.

"Anna," said Tom quietly, "bring me my shotgun from the closet and load it with shells you will find in the bag." Anna was a Valley girl and knew what was at stake. She quietly fetched the gun, loaded the shells, and slowly passed the weapon to her father who had not moved. The cat-and-snake standoff continued in the quiet afternoon, with neither backing off, both somehow locked in a deadly drama that did not look good for Bowser. Tom slowly and deliberately raised the gun.

It is said that rattlesnakes are so quick that they can strike at a bullet speeding toward them. Surely this snake never saw the load of buckshot that sailed over the cat and reduced the killer to snake parts in very short order. When the smoke cleared, Bowser had vanished. Tom had saved the cat's life but expected no feline thanks.

Later that day brought an extraordinary moment. As the family were sitting on the patio, Bowser came up to Tom and rubbed himself against his leg in an unmistakable and heretofore totally absent expression of affection. Tom found himself accepted and even loved by the cat from that day forward. He could hardly imagine that Bowser understood that his life had been saved and the explosive shotgun blast was meant for the snake and not the cat. But Tom accepted the new relationship with pleasure and never questioned the depth of cat understanding.

· *J* ·

Rat Fight

ALISON GREEN was the very talented winemaker of the Firestone Winery for twenty years. She had grown up in Sonoma County where her family grew grapes, and in addition to her connoisseur's enological skills, could work the hoses and move wine barrels with the best of the winery crew. She was also a first-class pitcher on a women's Valley softball team, a skill that would come in very handy when the chips were down.

She lived by herself in an old nineteen-twenties frame farmhouse. Her only companion in the house was a smallish cat who came and went during the day, but usually curled up on Alison's bed for a quiet, comfortable night. The gray creature, full of affection and personality, was known simply as Cat.

The country house was quiet and peaceful with one exception. Often Alison and Cat's nights were interrupted by the busy noises of rats in the hollow wall spaces and ceiling. Alison did not begrudge the rats their turf and livelihood, and her country upbringing caused her no concern about the rodents that kept pretty much to themselves and would never invade her refrigerator or other privacy. Also, there was Cat to stand guard.

The creatures continued to play rat games in the walls, building nests, moving things about, scurrying, and carrying out their territorial rivalries, sometimes with rat squeaks. The noises were muffled and only occasionally disturbed the quiet Valley nights enough to wake Alison or Cat, who mostly slept quietly through the noises. But once, in the middle of the night, the rat rustlings were more insistent than usual and both sleepers awoke with the realization that there was something moving about in the room. When the lights were turned on, a very large rat scurried from under the bed!

With a hiss and growl of rage, Cat was after the rodent in a flash, and the two came to a standoff in a corner of the room. The rat was trapped by the cat and rose in a rage on its haunches like a dinosaur, front legs flailing and evil incisors gnashing and chomping the air. The rat presented a terrible aspect; and the cat, equally enraged and with teeth bared and claws at the ready, held the rat in place but did not seem inclined to close into combat. The moment must only have been a couple of seconds, but in Alison's mind the picture was a long freeze-frame of animal rage and faceoff. The two ferocious and madly threatening creatures held each other in place with mutual challenge and threat of mayhem.

Alison was also on the fight with a desire to protect her cat friend and a thorough rat repugnance. Close to hand were her heavy winery work boots, and she wielded one with all her pitcher's might, felling the rat with dead-on aim. The rat was down and stunned, and in her own moment of abandoned rage, Alison beat the rat with the other boot in the heat of fear, anger, protection, and deadly accuracy. The rat was done in three whacks.

The rat was shoveled out into the night, and Cat preened and settled back, curled up again on the bed. To both Alison and the Cat's credit, they spent the remainder of the night in blissful slumber. The house was abandoned to the exterminator people for the next few days, who somehow convinced the rat population that there were better homes elsewhere for the remaining years of Alison's and Cat's happy life in the Valley.

· 𝒥 ·

Trouble Dog

MOST DOGS try to fit in with their families, sometimes with a temporary problem like occasionally wetting the carpet or barking at night, but generally dogs get along. The prize for the exceptional all-time loser dog in the Valley would probably go to Digger.

Digger arrived at the Adam Firestone family as a Jack Russell puppy looking for a home and nourishment. He soon grew up to be an energetic, intelligent, long-legged white and brown Jack Russell of unusual vigor and capacity for trouble. Like some Jack Russells, he could not sit still for a minute. In his early years he ate a couple of cats that had found their way to the household. He barked at any living thing that moved on the property, and when that was not enough, he tried to take over all the neighbors' yards.

Most dogs will adjust to a home territory and accept the limitations of roving. Digger could not be contained in any yard or by any fence. He would go over, under, or through barriers or, in a moment of carelessness, out any gate opening. He would run away to a selection of neighbors, who initially welcomed him, then soon learned his ways and called for him to be taken home.

Digger would chew furniture legs or fabric and, if nature called, would not wait for the outdoors like a normal house dog. If a strange animal happened by in the night, the entire family would be awakened. If the mood took him, he would bark to be let out or bark to be let in. He continually scratched and fretted. At length, his abusive ways were becoming too much to handle.

Of course the dog had no fear or judgment about taking on other creatures. One day, his barking became a roar, then panic, and then abruptly stopped. Adam's instincts were aroused, and he grabbed a gun and went outside to see Digger being taken away in the jaws of a large coyote. Digger would go after anything, but now he had met his match and was off to a coyote clan picnic! Adam fired into the air, and the sound startled the coyote into loosening his hold and, for once, Digger retreated quietly under the porch. On reflection, Adam wondered if he had aimed correctly or if he should have fired at all.

As time went on, the irritations grew, and the idea of somehow getting rid of the family pest began to dawn on the household. Deliverance came in the form of a carpenter who, in the course of building a garage, had taken a liking to the dog. The routine of sharing the carpenter's lunch in the front seat of his pickup probably had something to do with the bonding, but it was obvious that the two had a friendship. Finally the carpenter asked if he could have the dog. Adam asked no questions and Digger had a new family and, hopefully, a new and more uneventful life.

A story like this would probably end at this point, trusting that somehow a reformed dog and loving family would live happily ever after. This saga had a peculiar twist. Over a year later, a welder pulled up to the winery and got out from his pickup to talk to Adam. As they were discussing a welding job, Adam became conscious of a dog barking to get out of the truck. On closer inspection the dog seemed familiar and it was, in fact, Digger. "Who is the dog?" Adam asked innocently. The welder replied, "Somebody gave me that Jack Russell and I'd like to find a home for him! Are you interested?"

Adam replied, "No thank you!"

· 🦜 ·

Parrot Peregrinations

JOEL AND Charlotte Baker were returning home with their teenage son Zack, driving almost to the top of their driveway, when they noticed a mysterious presence at the side of the road. Inspection revealed a large docile parrot under a bush eyeing the Bakers as only parrots can do with an intense, sideways, one-eye stare. The bird did not seem alarmed or inclined to move, but appeared distinctly out of place. The big-eyed bird sported bright turquoise and yellow feathers, a gorgeous sight to behold, sitting alone and foreign to the Valley countryside.

The Baker-bird meeting was on a Sunday afternoon, and parents wondered what to do while Zack took the bird water and fruit, demonstrating an unerring instinct of a young person in sympathy with an animal. The parrot was hungry and thirsty and left his bush for the food and water. Next began a long process of tempting the bird to move, using more treats to chum the new friend along the road toward the Baker home. Finally bird and boy were close to the barn, and Zack actually put on a glove and persuaded the parrot to perch on his wrist and enter an enclosed room in the barn where, at least, he would remain safe until events developed.

This was an impressive opening act of animal recovery,

but what next? The Bakers phoned around town and finally reached the proprietor of a pet store who said, yes, he had heard of a missing parrot, and would initiate a search among bird owners from the store the next day, Monday. Meanwhile, the bird lodged overnight in parrot splendor on a rail in the Baker's barn, easily making friends with the sympathetic Zack, who has always had a knack with animals. By now the bird was full of fruit and water and inclined to be dozy.

The next day a surprised and grateful owner, Dr. Jean Seamount, was tracked down by the pet store, and the parrot identified as "Tiki." Parrots and owners have long-term relationships, and an open window and a moment of confusion caused the bird to leave home, travel around Buellton, and come to rest beside the Bakers' driveway. An immensely relieved Dr. Seamount retrieved the parrot, who was equally happy to leave barn and boy for a familiar home. The Baker family basked in the glow of a truly good deed successfully accomplished and a happy memory of the exotic visitor.

A year later, Zack needed braces and on good recommendation, headed off to a skilled orthodontist, whom the Bakers recognized from the previous parrot episode. Dr. Seamount immediately recognized her parrot's rescuer and was happy to fit Zack with the necessary braces that resulted in a great smile that will be with him the rest of his life. When the time came to settle the fee, the Bakers discovered that this work was a gift from the good doctor, with all gratitude for the shelter and return of her wandering parrot friend.

Absent Friends

ALL OF us who share our homes with animal friends live in fear that they will somehow become lost. Posted notices around Valley bulletin boards and post offices too often display plaintive pictures and stories of lost pets. The County Animal Shelter, Humane Society, and veterinary clinics have effected numerous reunions between distraught families and wayward family critters who have been turned in by good people or captured as strays. Every agency advises tags with phone numbers, but identification sometimes becomes lost. Who of us has not experienced those anxieties and dramas when our dog or cat friends do the natural thing and explore the countryside?

An all-too-typical saga began with a late-night party, a gate left open, and two family Labradors taking to the streets. The Allen family, Val and Matt and three children, loved their mature Golden Lab, Cannon, and the rambunctious two-year-old black Lab, Bowie. But the pull of the open road had brought the two dogs out of their Los Olivos home to the beckoning streets of the little town, and they were nowhere to be found.

Valley people are very kind and savvy about strays, and the people at R Country Market in Los Olivos called the number on the tag that same night to report that a dog named Cannon, tied up at the store and looking sheepish, was ready for home companionship. One was restored, but the other Lab was still on the loose somewhere with no news. The next day the search began in earnest, with parents driving the Valley roads and all the likely Bowie haunts. The Allen kids bicycled the town, calling and peering into backyards. There was no trace of the friendly young Lab, and, worse yet, the dog tag had come off and was found at home. The search continued and photographs and notices were posted around town.

Meanwhile, as the story is now known, Bowie had fallen into very sympathetic company. A certain young waiter at Mattei's Tavern had found the dog when going home late on the fateful night. Single men living alone and dogs looking for company can become instant friends, and waiter and dog had struck up a great relationship. The dog now lived in a bachelor pad on a ranch, rode to and from work in a pickup truck, slept on the young man's bed, and, best of all, lived on the very excellent scraps and leftovers from Mattei's that the waiter had in abundance. The waiter knew that the dog was probably a temporary friend, but did not have enough information to research the dog's real home.

Fortunately, a waitress at Mattei's saw the photograph and lost dog notice in the Los Olivos Post Office and suggested to the waiter that he would do well to call the owners. The contact was made and the five Allens immediately drove to recover their lost friend. Bowie was delighted to be back in Allen family affection and security, even if it was only in the backyard and on leash walks. He is still happy to this day, although, if the truth were told, occasionally when the breeze is right, the scent of the Mattei's kitchen wafts across town to the Allen's yard and Bowie dreams of the lavish bachelor adventure.

The story ends happily as most lost friend sagas do, although a few sadly do not. The Valley owes a debt of gratitude to the pros in the County authorities, Humane Society, and veterinary clinics who stand ready to help, and also all those good people who respond to a notice of a lost animal. May the Valley always be friendly to absent animals; and, again, thanks to all who help return the strays.

· ℱ ·

Barnyard Jungle

SOMETIMES IT is hard to contemplate fully or accept the apparent cruelty of animals to animals that happens, even in our bucolic Valley. It is not really cruelty, because there is an absence of malice, but nevertheless, violence lurks. A spider will eat a fly; a mountain lion will take down a deer; a coyote will stalk an innocent house cat; and don't anyone step too near a rattlesnake. These realities exist and can be pondered with sadness and sympathy, but they are animal world realities and not our own.

Wolcott and Nancy Schley lived in Fredensborg Canyon with a few acres and an antique barn bedded with friendly old straw that made a perfect home for a semi-wild mother cat. The cat was a good mouse hunter and prospered on table scraps, but preferred not to have anything else to do with the

Schleys. The cat had obviously found another feline friend, as was soon demonstrated by a large litter of kittens. The big-eyed, furry creatures were adorable and playful in the straw, as the mother cat came and went, making a home for the young. The Schleys were happy to have more mousers and fed and played with the kittens from time to time.

But there was something amiss in the happy cat life, as every few days a kitten would mysteriously disappear from the family. Sometimes in the morning there would simply be one fewer in the litter, without trace or explanation. Who could be catnapping the innocent kittens? The cat family was too wild to be sheltered and only marginally a Schley responsibility, but still the mystery needed to be solved and the litter protected if possible. Wolcott was a country person, wise to the ways of animals, but he searched the barn for clues and trace of missing kitten without success.

Finally, he determined to solve the mystery with a stakeout. He armed himself with a thermos of coffee and a strong flashlight and settled down in the reaches of the barn to observe the cat action during the night. The diminished cat family settled in at the other end of the barn. After a time, he noticed the mother cat pad off in the dark night to do cat business, leaving the sleeping kittens. Time passed and the night was quiet until the stillness was broken by a ghostly swooshing sensation and then a plaintive, receding kitten noise.

The next day Wolcott announced the solution to inquiring neighbors: "Owl Airlines!"

· 🦉 ·

First Animals

DURING HIS term of office, when President Reagan came to his beloved Rancho del Cielo in the mountains overlooking the Valley, he was delighted to escape heavy matters and heavy people for an always too-short stay on the ranch. The President was a friend to his Valley animal pals

and longed for dog and horse conversations as an antidote for the affairs of state. He and Mrs. Reagan missed their ranch pets when on the Washington circuit and took great delight in catching up with dog affection on the ranch. The dogs knew this because life was never the same with only ranch hands and security types around, and they too missed their pal, the President.

After a few restorative days of a presidential visit to the ranch, the leave-taking defined the President-and-dog relationship. The ceremony of First Family departure was as presidential as possible, considering the dogs ran the show. In hushed preparation, the spit and polish of the Marine helicopter and the attention of the surrounding military and Secret Service people gave a certain tension and formality to the ranch helipad area. Protocol people managed the guests invited for the sendoff, who would wait in hushed anticipation as the moment approached. All the formalities involving security, awaiting crew, and busy staff of the presidential circumstance weighed heavily on the country setting, awaiting the arrival. And then up would drive the First Jeep, with a beaming President, the First Lady, the ranch manager driving, and the dogs.

On one particular departure day, Victory, Lucky, Freebo, and Millie were having their last hysterical fun moments with their family and they took over with abandon. They were friendly, uninhibited dogs, like the kind that live next door and without any presidential hang-ups. Staff and security tried in vain to establish protocol and order as the President and First Lady stepped out of the ranch Jeep and approached the guests, but the dogs took over, bounding up to everyone with tails wagging, weaving about the scene with joy. The President would greet the guests, pretending his dog friends were not there, and trying to ignore their enthusiasms. Finally the moment of departure arrived and the dogs were restrained back in the vehicle, the President stepped into the helicopter, blades turned, and the real world reasserted its relentless self. The dogs watched the ascending first flight as quiet returned to the country.

The first dogs serve functions other than Presidential therapy. Once, my son Andrew was on a school hike which passed the Rancho del Cielo gate when ranch manager Courtney Travis drove by with Lucky sitting in the truck. The President was in residence, and security was tight, but security gave way while dog and boys got to know each other. The young students learned to disagree with the President's policies, but they will always identify him with Lucky and feel better about the Administration.

Another time a group of demonstrators were at the gate and the protesters brought some children with them to the political theater. Once again, Lucky came out and world affairs took a back seat to dog-kid affairs, and the whatever political differences existed lost an edge.

Horse therapy was also an important part of the President's visits. He was a first-class rider, and he would insist that only one agent, John Barletta, ride with him and all other security must stay out of sight. The President saddled his own horse and sometimes led John on a fast ride, to the consternation of the security types who fretted about horse accidents, but there never was a mishap. The exercise and contact with horses did the President a world of good, and he came in from his rides a changed man. It did go a little far when Mexican President Portillo presented the Reagans with a brilliant horse named El Alamein who had a tendency to shy and whirl on the trail and was something of a menace for the President to ride. The Secret Service asked veterinarian Doug Herthel to try the horse and see if there truly was a potential danger. Doug went for a ride on El Alamein and concluded that the horse was not a safe mount for any cowboy, much less the President of the United States. Fortunately, the President had confidence in Doug's judgment and agreed to give away the beautiful and flawed horse.

The President finally wound up with a beautiful bay gelding named The Sergeant who, as they say, was bomb proof. One bad day the President's cinch came loose and the saddle went sideways, depositing the leader of the free world upside down under the horse. The Sergeant just stood still until many hands unraveled the problem and everyone heaved a sigh of relief. When the Rancho del Cielo finally sold, the President gave the horse to his veterinary friend Doug Herthel.

One favorite true story, told in John Barletta's book, *Riding With Reagan,* took place on a typical morning ride. As the President and John came around a corner, a large and menacing rattlesnake straddled the trail. There was a standoff, and finally the President said, "Shoot him!" John replied, "Are you sure, Mr. President?" "Just shoot him!" the President replied.

With one shot the snake lay dead on the path. "Well done!" the President exclaimed.

Of course with the gunshot, all hell broke loose on the ranch with security on full alert. John quickly radioed the incident to headquarters, but that did not prevent Secret Service and Military back-up dashing to the rescue with weapons drawn only to discover a sheepish President and a dead snake.

As they continued the ride, the President praised the accuracy of Barletta. He knew the Secret Service agents were good, but not good enough to plug a snake with one shot. John Barletta could never find the courage to tell the President that the first bullet loaded in his pistol chamber was actually buckshot that scattered over the trail and couldn't miss the snake. Terrorist threats notwithstanding, the greatest danger on the ranch outings was really rattlers, and John traveled loaded for snakes.

The Valley misses the President, who loved his ranch and was a true country neighbor in his relationship with animals.

· ℱ ·

Part Six
Prize Winners

Performance Partners

CHARLOTTE BREDAHL is a tall, gorgeous, blond Valley woman and there is also something about her that exudes a steady, unflappable focus. This latter attribute is a very good thing, because her champion dressage horse, Monsieur, is a great animal, and anything but steady.

In 1986, Charlotte went to Denmark (of course) from the Valley to search for a prospect to train to be a serious dressage competitor. Europe has a wealth of large, sturdy, warm-blooded horses. These animals, unlike the swift Thoroughbreds, gorgeous Arabs, or other specialized types, are cross-breeds, probably descended from warrior knights' battle horses, and tend to good size and heroic performances. Charlotte was fortunate to find a five-year-old, barely broke, but very handsome gelding with obvious potential at a small farm in the Danish countryside. She also found a friend for life and a true dressage champion.

Charlotte explains the character of her beloved and talented Monsieur as being that of a mentally slow-maturing and somewhat insecure animal. This is another way of saying that the horse, like many exceptional performers with great physical prowess, is a little nuts. He does have his very individual ways.

On returning together to the Valley, the first time Charlotte rode him he bucked her off in short order. He never did this to her again, although he certainly knew how, because, as Charlotte maintains, he had made his point and did not need to do it again. He destroyed every horse blanket he could set his teeth to and finally had to wear a special bib to protect blankets, leg bandages and all else from destruction. Once on a plane trip, before takeoff Monsieur destroyed the tack box and Charlotte's suitcase with his teeth, but this was after a few great wins when he could be forgiven.

On returning to the Valley from Denmark, Charlotte's focus and skills were called on as they began training in earnest. The discipline of dressage requires exceptional athletic ability from both horse and rider and complete control of movement and concentration. This exceptional horse needed to be trained slowly, and the two spent hours and long days gaining each other's confidence. Coaches and trainers were consulted and happy to work with Monsieur, but Charlotte was always the one to be aboard. Once the great trainer Herbert Rehhbin briefly tried the horse and advised Charlotte, "Don't ever fight with this horse and you be the one to ride him!" He recognized the horse's personality and told Charlotte that she and the horse had a special relationship, and that was that.

In 1990 Monsieur did exceptionally well in all the Olympic United States Equestrian Team qualifying rounds on the West Coast, but, in the final selection, it was agreed that the horse was too inexperienced for this demanding competition. Charlotte was asked to try again in two years.

Meanwhile, the two partners continued to win in dressage Grand Prix competitions. Originally, in partnership with the individual who had put up the money, the idea had been to complete intensive training, win awards, and then sell the horse, who would be very valuable when successful. The plan almost worked out, but it soon became evident that prospective buyers, even the best riders, did not get along with Monsieur as did Charlotte. The horse grew in reputation and success, but only as a partner.

The preliminary competition in the 1992 Olympic Trials, held in Orlando, Florida, ended in a disaster for the two. Next door to the show ring, a school had let out with crowds of yelling kids exploding just as Charlotte and Monsieur came into the ring. The horse lost his cool, and Charlotte had to improvise a routine in the half of the ring away from the kids. But fortunately there were more rounds and the two came back strongly in subsequent competitions and were selected to be members of the United States Equestrian Team for the Barcelona Olympics.

At Barcelona, Monsieur was in peak form, but a new distracting challenge awaited. As the two came into the competition arena for the Olympic finals, the huge crowd started clapping and our hero decided he wanted to leave the stadium in a hurry. Fortunately, the moment a horse enters the ring, the dressage competition has not yet officially begun and there is a safe period, without judging, when horse and rider settle down before the bell sounds. Normally the rider will pass by the judges' booth and salute during this time, but Charlotte was having a problem going by anything in the arena. However, Monsieur did have a very handy and reliable ability to back up, and Charlotte reversed course and backed the horse by the nonplussed judges, who accepted the unusual salute. Monsieur then put in the performance of his life and Charlotte won an Olympic Bronze Medal!

Accepting an award was always a problem for Monsieur. This would usually require standing still for the ceremony, sometimes for a band playing an anthem, often with flowers or blankets draped around the neck, and always accompanied by cheers and a victory lap. Monsieur hated this show and would have nothing to do with the award accolades. Usually, Charlotte rode in to accept her prize on a borrowed horse loaned by an understanding competitor who knew Monsieur's individual peculiarities. In Barcelona, a sedative was the only solution, and Charlotte explained her problem to sympathetic Olympic officials who granted a concession inasmuch as the competition was over. On this one occasion, a happy and docile Monsieur participated in the award ceremony and received his Bronze Medal with the help a strong dose of otherwise completely banned equine pharmaceuticals.

Charlotte and Monsieur live happily today on a ranch overlooking Buellton, and the two will occasionally participate in a demonstration event. There are a number of very good riders in the Valley on presentable horses, but nobody comes close to the class of these two when strutting their very talented stuff. The horse and all his horsey quirks is happy on the ranch and always will be. Charlotte generously helps young aspiring riders and the Valley is lucky to have the two living here.

· 𝒥 ·

Mule Days

THERE WAS a time in the '70s and '80s when mule racing was the passion of a number of ranchers in the Valley.

Mule racing is like old-fashioned county fair horse racing. The moody mules who might or might not want to run on a given day, possibly with a little pharmaceutical assistance, will be paraded out and ridden by amateur riders. The course will be a straight track of about 200 to 400 yards. The start will be "lap and tap" which means they will walk toward the starting line in more or less even order and, if the starter thinks the pack is reasonably fair to all, a hat will be dropped and they are on their way. Mules are more independent-minded than horses, and sometimes they run fast and sometimes they don't. The assembled crowd loves this, and greenbacks change hands freely without any parimutuel or racing commission. A mule that could run was highly prized and brought the owner cash and bragging rights.

In 1974, local ranchers Dean Brown, Joe Russell, and Art Clausen heard of a superior mule and formed a syndicate to travel to Bishop, California, to see her in the races. Dean particularly fell in love with the animal, and they gave $2,500 for the mule called Ruby, a good sum for a mule in those days. She won the big race in Bishop and returned to the Valley to begin her career. She won some local races, won the Rancheros Visitadores mule race three years in a row, and began to gain a reputation in the ranchers' underground mule circuit.

As mules go, Ruby was a beautiful animal, almost sixteen hands tall, and had endearing, friendly qualities. She had a slick coat and upright bearing and responded to the spoiling on Dean Brown's ranch with a reserved, mule-type affection. There was true personality inside the huge head.

The following year at Bishop, Ruby was entered with Dean's son, Lance Brown, as jockey. A new start system required the mule riders to hold on to an overhead rope end to ensure everyone set off evenly. Of course this meant one hand on the rope and only one hand on the reins, which led to steering problems. When the start was given, the independent Ruby responded by whirling around and facing backward toward the stable while all the others took off in the right direction. Lance finally got her going the right way down the track, and despite the late start, caught up and finished fourth in the big race.

One day Dean received a call from El Paso, Texas, that sounded something like this. "Halo! Weuns hear yuall got a fast ass. Well, weuns got one too an are wanten to put up twenny five hunner dollas if you all wanna bring yohs heer an see whot's whot." Dean knew of the boys in southern Texas and understood the language. The challenge was met and the day set. Ruby would take on the local champ mule in Texas!

A friend of Dean's syndicate had just taken delivery of a forty-foot, air conditioned, quilt-lined, chrome-wheeled, and bright red horse transporter. The plush Thoroughbred outfit was heading east without a load and it was agreed that Ruby would arrive in style. She was outfitted in a silk-bordered racing blanket with a bright new silver halter. Even her hooves were polished and she seemed to sense an important challenge as she proudly left for Texas.

The Texans laid out the track and prepared for the festivities. Most of the hosts owned huge ranches, and in Texas that is measured by square miles. Many of those square miles were over oil fields, so the festivities would be painted with a broad brush. But that is not to say that $2,500 and pride on the line was not a big deal.

The California syndicate traveled by a private plane supplied by another friend and landed at the runway of the Texas ranch on the appointed day, timed for the well-orchestrated arrival of Ruby. At precisely the right theatrical moment, the gleaming truck and trailer came down the road with chrome air horns blaring. Ruby rode the big rig in splendor, accompanied only by an old horse friend to keep everything calm. At the arrival, uniformed attendants sprang out to unload the champion mule and the Texans just looked as Ruby strutted down the ramp in her silk blanket with a true sense of occasion.

Despite the intimidating arrival, the Texas betting money flowed freely, only somewhat subdued by the grand entry. It is recorded that the Texas mule entry had a quick bath and ranch hands hastened to clean the tack although they could find no hoof polish. A Texan exclaimed "Yoh fancy pants ass is in fo a whoppen from our lil country mule so don keep putten lipstick on that pig." Fortunately the man also whipped out a role of hundreds to back up his words.

The afternoon match race was on and was over in a California hurry.

The two mules and jockeys paraded to words of encouragement and insult, lined up at the start and got off evenly. For once, when the starter dropped his hat, both mules ran straight and flat out with riders whipping and the small but intense crowd yelling. At first the race was even, but then Ruby's long stride did the job and she pulled away from the Texas mule by three lengths at the finish.

Dean soon had an extra $2,500 in his pocket in addition to all the side bets his pals had been able to promote from southern Texas pride. But Texans don't stay beaten for long and ranchers are ranchers and the festivities began under the big southwestern sunset.

Ruby departed in her splendid transportation and everyone began a true Texas big-time barbecue that lasted in grand style into the night. The California friends flew home the next morning and the celebration continued with accolades for the home team from a proud Valley. In her career, Ruby won over fifty thousand dollars for her owners. She finally retired to Joe Russell's pastures, gazing into the distance with a contented look. It is said that mules never forget, and there were a great host of memories in that mule brain.

· *J* ·

Polo Partners

THE FIRESTONE Walker Brewery in Buellton keeps monogrammed mugs for locals who frequent the restaurant, as is the custom of many famous pubs around the world. The form is to have one's name etched on a mug to enhance the comfort level while enjoying the brew. Joel Baker, one of the best athletes in the Valley and certainly the premier local polo player, is welcomed to the Brewery with his personal mug sporting the name "Hydraulic" on the glass instead of his own name. Only those who are strangers to Joel would ask why.

In the seventies, before moving to the Valley, Joel was a young stock broker and skilled polo player who frequented the old Will Rogers fields located on Sunset Boulevard, west of Beverly Hills. Important Hollywood characters would congregate there for fashionable polo, and a personable young athlete like Joel was popular in that he could really play the game and added to the fun.

Talented and athletic horses are everything to success in polo, and a person with a good string can outperform a better player with rank horses every time. In those days, a string of at least four polo ponies was required to participate, and that could be a financial strain on a young man. Joel had a limited-budget lineup of horseflesh but still did well. In the off season, his string would travel to the Valley to board at the Jack Smith Ranch near the Santa Ynez River. Jack was a successful Hollywood writer who purchased green horses with potential and then trained and sold them as polo ponies in partnership with his son, Andy.

One day Joel traveled to the Valley in the company of an inexperienced, high-roller polo player who was in the market for horses and wanted to look at the talent at the Smith Ranch. The man tried some horses, including a young and only half-trained small Thoroughbred, with very little success. Here, Joel earned every bit of satisfaction and great memories received whenever he looks into his beer mug, because something caught his attention about the way that horse moved, of all the horses they tried that day, a talent that the buyer and the Smiths missed seeing.

The horse was a young fifteen-one-hand black Thoroughbred that had been bought off the track by the Smiths and given some polo training but really not much attention. The horse was named "Hail to Howard" and had a poor track record. His front feet toed in and he had a paddle in his gait that caused the front feet to kick out sideways. The horse did not seem to respond well on the field, and another problem was that Andy Smith could not catch him in the paddock. After a little horse trading, Joel bought the animal for a bargain price and the Smiths thought they had the best of the deal.

When time came to leave the ranch, Joel found that he had no trouble in the paddock putting a halter on the black Thoroughbred, and they somehow got along together from the start. The first time the two played in a practice game at Will Rogers there seemed to be a bond. Before long, the little horse became the first choice of his string, and they tended to win games together. Joel for some reason renamed the horse "Hydraulic," a name that would soon be known on the circuit, and together they grew into a winning partnership.

Thoroughbred horses are bred to compete, but usually by way of running straight and fast. Polo has a good deal of running, as the horses charge down the field after the ball, but stopping handily and turning are equally important, and most Thoroughbreds have a hard time understanding this. Sometimes horses need to run and bump each other off the line of the ball, and that involves pushing and shoving, sometimes at full speed, which will cause panic in horse minds that don't like this interference. The thrill of competition when playing polo occurs to only certain great ponies, and Hydraulic understood and appreciated the game completely. Additionally, every horseman knows that individual people do well on certain horses where someone else cannot get the job done. Hydraulic responded to Joel for some horsey reason and the two were a team.

Hydraulic placed Joel on the ball fast and true, and Joel was only required to power the ball down the field to the right

place, which he was very able to do. If running and bumping were needed, Hydraulic always gave better than he got. Together they were magic on the field.

Even great horses seldom last at full play for many years, but Hydraulic actually played twenty-three successful seasons. At his peak form in 1981, when Joel was a seven-goal handicap (which is pretty good for a financial advisor), Joel was on the winning United States Open team at the Nationals in Retama, Texas. Here, Joel and Hydraulic played on a twenty-six-goal team in the most important tournament in the country. A polo game lasts six exhausting periods, and a fresh horse is usually played each time. In the final match of the tournament, Joel played Hydraulic in the second and sixth periods and the little horse delivered 110 percent each time as the team won the National Championship.

Hydraulic carried Joel to win four Pacific Coast Open tournaments. Every time they played together, Joel had the confidence of knowing that he had an exceptional athlete to carry him and an animal that loved the game as much as he did. The little black horse with a great heart and a love of the game became famous.

Finally, Hydraulic retired to the bucolic horse pleasures of Joel's Buellton polo ranch. In the horse's retirement years, sometimes Joel would take him along to the Santa Barbara Polo field, where Hydraulic would watch the game from the sidelines. The horse loved polo and actually followed the play and would mooch up to a mount that had just finished playing a period, when the horse came in hot and tired. Hydraulic was no usual horse. He understood the game that he had played for so many years and, unlike most dozy horses on the sidelines, he pursued his love of the game into becoming a spectator.

Hydraulic finally passed to the fields in the beyond at the age of thirty-six. This was one of the superior animals of all time. He has a monument on the ranch and a place in Joel's heart forever. Hydraulic's name is on a mug in the brewery, and we can forgive Joel for staring in the brew occasionally with a host of great memories.

· *𝒥* ·

The Champ

ALTAMAS' MYSTIQUE, the ultimate German Shepherd, only rates as a Valley Animal because she graced the Lompoc Valley Dog Show to strut her stuff and stayed in a Buellton Motel.

The Lompoc Show is a big deal for the community and an officially sanctioned dog competition that gives winning entrants points that are important for the National Championship. Lompoc also provides a very pleasant country experience that has all the prestige but less of the hassle of most high-energy urban and more pressured dog shows.

My aunt, Jane Firestone, married to my Uncle Ray, who was retired Chair of the Firestone Tire Company, had a passion for German Shepherds, and a very keen sense for choosing and campaigning a potential champion. Her first great success had been named Manhattan, a male German Shepherd who became a United States Champion after cleaning up most of the shows in the country. Her present campaigner, Mystique, a female, had similar and even superior qualities, and in time won Best of Show an amazing 275 times for an all-time record of Bests.

At the time of her Valley visit in 1983, Mystique had won around a hundred Best of Shows, and was a highly acclaimed competitive dog slated for greatness. She traveled first class with her handler, Jimmy Moses, Jimmy's wife and show partner, and the Champ's best friend, a Welsh Corgi that was worth a few prizes himself. The traveling troupe arrived a couple of days before the Lompoc show to breath the local air from a Buellton motel and prepare for another championship. My uncle and his dog impresaria wife Jane arrived on the day of the show in similar first-class mode, staying with friends in Santa Barbara.

Kate and I were invited to watch the excitement on the show day, and exciting it was. The preparation, attention to dog turn-out detail, and precise minute-by-minute staging of a champion dog is an event to behold. Mystique's show day began with a quiet early morning walk, followed by a bath and coiffure worthy of a queen. She was exercised, cleaned, brushed, and pooped; her eyes and nostrils were scoured; and she enjoyed a brief training table meal. The entire purposeful routine clearly conveyed to the dog mind that this was to be a big day.

The show began and our champ entered the ring to be judged in a succession of classes including female German Shepherds, then males and females together, then all ages, then winners of all the Shepherds, then the Division, and so on up the line to Best of Show Class. In each competition, the dogs would trot into the ring, line up for inspection, present themselves individually for a pat-down close inspection, then trot out before the judge, who put them in order for group inspection and carried out further drills to choose the class winner. The sweepstakes prize came at the end of the day, when a judge selected the premier animal of all the entries. Time after time Mystique was led into the ring to be judged, and each time she seemed to make that moment the most important of the day. She began to win steadily up the ladder.

And here we learned that there is more to showing than just a good shampoo and trot-out. The handler's showing skills, relationship with the dog, and quality of presentation are all of utmost importance. Of course there must be a very good-looking dog with great conformation, but that is just the beginning. Consider a lineup of dogs, all clean and bright and all looking good with earnest and attentive handlers. A certain presence and bearing enables a dog to stand out and be a champion.

Clearly, Mystique was exceptional and looked the part among rivals. Her coat shone and her eyes gleamed. She stood tall and proud and had the champion look. Mystique had an extra something that seemed to say to the judge, "Look at me, I am special!" But there was even more to the presentation than met the eye.

The day was crowded and the show rings were circled with spectators and, of course, dogs everywhere, carried or on leashes and creating dog confusion. Jimmy Moses's wife had conveniently brought Mystique's Corgi friend to the show. His very presence calmed and reassured Mystique and gave her a devoted traveling pal in the long periods out of the ring waiting for the next class. When she went off for a judging, the Corgi seemed to say, "Good luck, Champ!"

But I soon noticed that when Mystique joined the competition in the ring, Moses's wife would take the Corgi in her arms and stand outside the area, strategically placed and very attentive. After the initial trot around, the dogs would line up for the judge to look at each in turn. After a critical glance down the line to size up the group, there was a close inspection, one by one, to rate the competing animals. The first group look was critical and that was the important Corgi cue.

Just as the judge strode out to observe the line of dogs, the Corgi would give out a cheerful bark, of course on the command of Moses's wife, just out of sight in the crowd. In the show confusion the yip was hardly distinguishable, except to Mystique, who knew well that it was her best pal calling, and her stature and ears perked up, searching to find the Corgi. It was a beautiful sight to see Mystique straining and poising to look for her Corgi friend, and it clearly set her out from the others in the lineup. The winner was usually a done deal with the judges' first comparative look and Mystique took the silver while the Corgi melted back into the crowd.

Jane Firestone looked great in the Best of Show photograph with Mystique, Moses, judges, and show Chair. Another win was in the bag, and the beautiful dog had done her thing once more. Kate and I found out another dimension of the show when we offered to send the trophies back east to my aunt's home. The show haul included about eight big cups and other sundry odd-shaped prizes, and the shipping cost me a fortune.

· *F* ·

Future Husbandry

ANYONE WHO worries about the future of our young people needs to go to the Santa Barbara County Fair and have their optimism batteries charged. The 4-H and Future Farmers of America (FFA) barns are bursting with the greatest kids in the County, who are busy concentrating on their club programs of raising animals and showing and selling them at the Fair.

Some 1300 animals are squeezed into pens in the Junior Livestock program Fair barns. Dairy cattle, goats, swine, sheep, chickens, rabbits, and turkeys have been selected at an early age and raised by 4-H and FFAers to be presented this big time once a year. The young people hope to win conformation and showmanship prizes and sell their animals at auction to reap the rewards of many hours of animal care. Many adult volunteers organize the formidable logistics of all these activities. The junior livestock event is an impressive and infinitely rewarding week of hard work and fun.

A talented and typical pair of program contestants are Allan and Carlene Jones's twin daughters, Madison and Clara, age eleven. They come by their Valley farming heritage honestly through their grandfather, George Jones, who operated the Pork Palace Ranch on 101 south of Buellton, and their father and uncle who were successful 4-H entrants as teenagers. The twins were well prepared to succeed and enjoy the Fair to the fullest as members of the Los Olivos "Lucky Clover" 4-H program and chose goats for their individual projects.

The goats, named Hamlet and Turkey, were purchased in early March and selected for conformation and disposition, although goats are moody and notoriously misbehaving and difficult to predict. The girls would attend weekly Lucky Clover meetings supervised by their advisor, Robin Flynn. They worked hard with their goats, providing nutritious feeding and grooming to ensure a healthy and good looking turnout, and additional hours of handling and leading work would develop a well-behaved presentation in the Fair. Hamlet and Turkey, handsome Boer goats, grew and gradually learned to be good 4-H citizens, enjoying generous and friendly care from the twins.

A very goatlike departure from the program was initiated by Hamlet, a notorious climber and escaper, who led Turkey out of the pen and Jones's yard to enjoy the freedom and delicacies of neighbor Orin Jensen's garden. Unfortunately, this coincided with Orin's effort to sell his house, and the prospective buyers were somewhat put off by wandering goats in the yard. Allan calmed his neighbor with an apology from the girls and a bottle of "Flying Goat" brand Cabernet.

Finally in mid-July the 4-H and FFA clubs convened at the fairgrounds for the culmination of the projects. Madison and Clara joined their cousins in a large motor home in the Fair campground, surrounded by hundreds of other families. They would be with their Lucky Clover pals, enjoying the company and Fair atmosphere, and parents would barbecue every night. Their goats were penned in the crowded barns, along with the hundreds of other animals.

The first challenge was on the Tuesday for the market-ready class judging, concentrating on the conformation of the animal. An intense beauty parlor treatment prepared Hamlet and Turkey to shine for this coming-out occasion. Madison won a fourth place and Clara an eighth place ribbon which was quite respectable in the company of many good looking goats.

The judge was a young man imported from Texas, Mr. Tye Fowler. He had grown up on a family goat farm, studied animal husbandry at Texas A&M, and passed an examination for Boer goat judging. He was very much in control, enjoying himself, and had won the respect of the parents and volunteers. There was much nervous tension in the judging ring, with well turned-out 4-H and FFAers focused and proud of their entries, and ringside parents, project leaders, and volunteers closely following the competition.

The next test came Wednesday with the showmanship class. Here, judging is on the young people and not on their animals. The general idea is that the entrant should be composed, focused, and confident, somehow compelling the judge to notice the show animal. As Madison and Clara trooped into the ring with other contestants, one could notice small nuances of presentation and animal handling that separated out the better contestants and gave the advantage to certain entries. The judge gradually selected those to be excused from top prize selection in reverse order, and the excitement mounted. The least winning would be excused to a second line, until only two were left for first and second. Clara was in the respectable middle of her class with a sixth place, and Madison brought the family a thrill when she lasted in the line up to place third in her class.

The final day, Saturday, is auction day. The first order of business is a lavish barbecue for invited buyers from the County who generously support 4-H and FFA. Anyone who has never attended auction day should seriously consider joining the excitement and supporting the splendid young people. The ring is surrounded with families, friends, and numerous volunteers, all hoping the kids will get a good price for their work.

One by one, the young people lead their animals into the charged atmosphere of the ring for an exciting and professional auction of their animals. Volunteers will serve as stewards to keep the show moving along, energetic and demanding spotters will work up the excitement and intensity of the crowd, and a professional auctioneer will call the bidding. When a purchase is made, the buyer will indicate whether the animal is to be sold on the commercial market, with the normal price being deducted from the auction price; or whether the buyer wants the animal to go to a processor, from whom the buyer will pick up the meat for home consumption. Of course the 4-H and FFAers know from the start of the program that this is the destination of their project animals.

The economic realities of 2009 brought concerns to the Fair as to whether the usual financial support would bring a good auction for the kids. The sale day began at eight A.M. and ran straight through to nine-thirty P.M., and fortunately all the animals brought even better than usual prices. Madison and Clara, into the ring after six o'clock with their goats, did very well on their sales. The Jones family is large and well known and additionally, doting grandparents were in the audience to support the twins. To add to the bidding, an outside buyer entered the auction at the last moment and made the day for the twins. It is recorded that The Firestone Walker Brewery took an intense interest in goat programs at the 2009 Fair!

· 𝒥 ·

Fancy Flying

IT IS difficult to imagine a more obscure and esoteric, yet more keenly pursued, sport than pigeon racing. Hardly anyone in the Valley, beyond the few who maintain a flock of the amazing birds, knows that a given sunny day may include an intense competition in the skies. But the eight or so flocks that live here are managed by fiercely competitive neighbors who maintain the birds and are engaged in racing as enthusiastically as entrants in the Kentucky Derby. The racers like Ken Hunter just go about their business a little more quietly.

A typical race will begin some four to five hundred miles away from the Valley, when pigeons are released out of a specially designed truck that has driven the precious racing pigeons to the starting location. The driver will rendezvous at the release point to which similar truckloads of birds from lofts all over the country have been driven and await the starting moment. At the signal, cages will be opened, the flock of racers will circle for orientation, and the pigeons begin their long, lonely, and very purposeful journey home. Later that day, and after sometimes as long as ten hours of fast flight, the pigeons will arrive at their own home loft. Pigeons fly from dawn to dusk and can average fifty to seventy-five miles an hour all day long.

When the birds pass the entrance to their loft, they will cross a device that transmits, from the chip affixed to their legs, the precise moment of arrival. A central computer will determine the trip speed in yards per minute and the race headquarters location will determine winners. The results will then be published to the eagerly waiting bird owners who

celebrate at home in lonely satisfaction, or in local pigeon racing clubs such as we have in the Valley. This is not everyone's
sport, but Valley neighbors who are engaged are as enthusiastic as bullfighters in Spain, steeplechasers in Ireland, or polo
players in Argentina.

Flocks are maintained in backyard pigeon lofts, where the
racers are pampered as only champs will be. They are purposefully bred, pedigreed, and perhaps crossed with stock
from Belgium, where pigeon racing is a national sport. The
birds come in many colors, from pure white to spotted grays
and reds, and eye colors range from yellow to blue-gray. They
strut, coo, and ruffle in their coops, enjoying a training table
any athlete would envy and pure, fresh water. The birds are
all carefully raised and maintained with a well-planned training schedule.

The conditioning program for a young bird will begin with
a short trip of about ten miles from the coop. Rarely will even
a new bird fail to reach home, guided by a mysterious "mag-
netic patch" as described by owners who readily admit that
nobody has ever discovered the true source of a pigeon's extraordinary ability. Studies have revealed that the birds are intelligent, with acute hearing and superior eyesight. They are
also great athletes, with long endurance and ability to outfly
and outwit pursuing hawks or to penetrate storms, fog, and
wind to make their way home loyally to nest and families.

Pigeons will cover extraordinary distances, and the breeding, care, and training determines how fast. Homing pigeons
were used by the Romans and throughout all history for carrying important economic and military messages, as well as
for racing. The breeding of the Valley birds may have originated between the Tigris and Euphrates, and the enthusiasts
are acutely aware of the heritage and historic lineage of their
sport. Our Valley breeders admire and love the birds and have
endless stories of their amazing prowess.

David Hunsicker has maintained a loft for years in the
Valley and enjoyed considerable success in racing and breeding. A few years back he brought glory to local pigeon flocks
by winning one of the most important pigeon races in the

country, the San Fernando Valley Club-sponsored Snow-bird race. The Snowbird could expect over two thousand pigeon entrants and would cover around four hundred miles, from Marysville to various home coops in the San Fernando Valley.

Many months before the big race, Dave brought a young hen that showed real potential to a pigeon trainer in the San Fernando Valley. The young bird, named only the "60 Hen" after the number on her leg band, demonstrated an excellent homing talent, speed, and consistency that matured into a likely contender for the big race. The hen was the product of many years of selective breeding and now it was up to the trainer to condition her for the big race. The trainer's coop had become her home.

For the Snowbird Race, the 60 Hen had been trucked to Marysville well before the race in a special van with many other club birds. She was fed and watered in the van, which was parked waiting for the signal in a field near Marysville. At the first-light starting signal, similar vehicles released the many birds, who then circled and sorted themselves out to make the long trek home. Meanwhile, Dave drove down from the Valley to the trainer's loft to await his prize pigeon's arrival, along with a number of the others racing from all over the country. Flying time can be delayed by wind, fog, storms, or predators, but the day was clear and fast and the birds were expected mid-afternoon.

Forty pigeon owners from the trainer's loft eagerly awaited the racers' arrival. Finally a bird was seen circling above, and it soon alighted and entered the coop. It looked like the 60 Hen, and Dave held his breath while the trainer quickly removed the racing leg band and stamped it in the time clock. The others cheered when Dave discovered she was indeed first in at that location. The 60 Hen, beautiful and courageous as she was, enjoyed a long drink and some food, then hopped to her nest where she sat on her eggs.

Dave proceeded to the next stage, race central at the San Fernando Valley Club, where the tension mounted as the crowd awaited race results tabulation. To add interest, the

City of Hope Hospital sponsors a betting pool, a percentage of which amounts to substantial hospital support. The odds are always long, but Dave had made a large bet on the 60 Hen, knowing that he had a strong contender. When all times were in from Southern California, he found out that the 60 Hen had indeed won, the trophy was his, and he took home 47,000 dollars!

Our hero pigeon has since been the dam of numerous descendants who have brought glory to the Hunsicker flock. However, unlike the recognition that attends the success of some of our local champions, most of us are unaware of our pigeon winners. Next time we look up and see a Valley pigeon making a straight flight for home, let's give a cheer.

· 𝒥 ·

Horse Talk

ENGLISH BRIDE Kate and I have been married for over fifty years, which makes me something of an authority on British character. They are traditional but not dumb, and that is the reason why when Sir John Miller, Equerry to the Queen of England, heard about Monty Roberts, he decided to investigate his methods of horse training. Sir John was a personal friend of a Valley character, John Bowles, who knew Monty well. When Sir John visited John Bowles in the Valley, he made a point of meeting and observing the horseman he had heard about. He was so impressed that he invited Monty to visit England and demonstrate his concepts to the Queen.

Monty, a long-time Valley breeder and trainer, had achieved a gentle and sympathetic method of breaking and training horses, as opposed to more harsh and domineering traditional methods. Everyone knows of the classic cowboy showdown in the corral when a man struggles with a horse to get a saddle on and ride, and where the word "breaking" has a literal connotation. Monty Roberts developed a far gentler means to Join Up® horse and man through communication and persuasion instead of physical mastery, and has enjoyed considerable success.

In England, Monty was ushered into the Royal Stables riding ring at Windsor Castle to train and place a rider on a young, totally green, and unbroken horse in the presence of the Royal Family. His success with the youngster in accomplishing this impressed the Queen, a considerable horsewoman and racehorse owner, who determined to spread the word about Monty.

Even with her support, the traditional English horse people viewed someone whom they saw as a Santa Ynez Valley (wherever that was!) cowboy with some suspicion. They were hard to convince and tried to frustrate his training approach by giving him rank and difficult horses for his demonstrations. But Monty prevailed and continued to deal with the untrained colts and fillies and mount his rider far more successfully than the onlookers could have imagined. At length he won Royal sponsorship and widespread English approval, and his concepts and reputation advanced considerably.

It is said that a prophet is never recognized in his own neighborhood. The California horse establishment had taken interest in Monty's methods, but initially were not all that impressed. However, with the English publicity, locals began to take notice and the training of horses took an historical turn. A thorough review of the old-style training process has now come to convince the horse world that there is far more sensitivity and perception needed in approaching a horse than was ever thought before. Monty Roberts's methods are now famous and, from our Valley, he has made history in the horse world.

Monty has lived with horses from early days as a Salinas cowboy kid, and has progressed to a world-famous race horse breeder and trainer, rodeo winner, show horse winner, author, and theorist. The path was not always easy and there were hard and impecunious days of discouragement before his present success. The one enduring ingredient of his life, in addition to the total and talented support of his wife, Pat, has been a love and understanding of horses. There are countless stories about Monty, but perhaps his life with Johnny Tivio best explains him.

In 1960 Monty was on the show circuit in California and took strong notice of a blood bay Quarter Horse stallion named Johnny Tivio that always gave him tough competition even though the horse had less than the best training and care. Monty knew the only way he could beat the horse was through the carelessness of the other trainer. The saga here takes a turn out of Western novels, as the horse was mistreated, taken out of competition, and in poor health. Monty was able to buy the horse that might never compete again for a bargain price, believing he could work with the animal. Miraculously, two years' nurturing and gentle training brought the horse back to his great potential, and now he had a rider that understood his exceptional talents. Monty summed up his relationship with the horse by saying he needed a request for performance rather than a demand, and love instead of neglect. Finally, in 1964, Johnny Tivio was ready to be shown.

The Salinas Stock Horse Show was in Monty's original home town and one of the most important competitions on the circuit. Monty consulted with Pat about whether to enter Johnny Tivio in the cutting or the reined cow horse division and the two could not decide. Finally they took the totally unprecedented but technically legal course of entering him in both divisions! The show committee came unglued and friends warned him against the plan because they did not want him to embarrass himself or tarnish the potential image of the horse. Cutting and reining are totally different disciplines and call for radically different approaches to handling cattle that are part of the class.

Finally the Robertses went ahead and made the dual entry, to the amazement of the competition and show-goers. The amazement continued when Johnny Tivio put in stellar performances and won both divisions! The rules are now changed to prevent any "embarrassments" like this again so it will go down in the records that he was the only horse ever to accomplish this dual feat.

The great horse went on to win four world championships and later stand at stud at the Flag Is Up Farm on Highway 246 between Solvang and Buellton. Alas, we outlive our horses, and he is now buried on the farm where his many memories live on. Monty is still training people and horses on the farm and, of all Valley animal people, has made history in his field. Training and riding are changing because of him, as horse people focus and concentrate on talking horse. Anyone who has an opportunity to visit the Flag Is Up Farm and observe Monty Roberts in action has a rare experience.

· *F* ·

Cow Love

TODAY, DR. Doug Herthel is one of the preeminent veterinarians on the West Coast, with a famous clinic employing fourteen veterinarians and performing equine surgery on everything from ranch horses to precious Thoroughbreds. But that was not always the case. In the early seventies, when Doug first moved to the Valley, his clinic was a two-horse barn across from where St. Mark's Church in Los Olivos now stands, and he and his competent and beautiful wife, Sue, performed operations on the lawn, with Sue administering the anesthetic.

The founder of the Alamo Pintado Equine Clinic may be a great vet, but he has his soft spots. In the mid 1980's, Doug was called to Louisville to consult on some Thoroughbred horse problems. During his visit, he and Sue, with their then four-year-old son, Mark, attended the Kentucky State Fair, wandered into the cow barn, and fell in love with a young Jersey cow that had just won the Dairy Cow Championship. Doug still will not divulge what he paid for the animal, but he bought her on the spot, and the cow, named "Nancy T," rode in first-class race horse transportation to the Herthel home in the Valley. The girlfriend of the transport driver performed the milking duties as needed during the trip.

They had not intended to go into the dairy business, or even move a cow to their home for whatever reason, but Doug is a keen judge of animal character and recognized something truly exceptional in the cow. Sue and Mark just fell for Nancy T. On arrival at her new home she moved, as contented cows do, to a lush paddock of green grass and delicious feed supplemented by the then beginnings of the Platinum Performance special formula. At first there were milking duties to be accomplished, but the Herthel family doted on the cow and no task was too onerous. While the milk supply lasted, there was more than enough, even proving that it is possible to have too much homemade ice cream in a family with two growing boys.

Nancy T in her turn responded with personality and affection that bore out all Doug's animal perception when he first met her. She communicated warm cow thoughts and pleasant domestic happiness to all around her. Some families include dogs or cats, or birds or hamsters or whatever. The Herthel family included a cow.

Somehow the authorities of the Orange County Fair heard about the exceptional cow and traveled to Los Olivos to evaluate the animal as a "Symbol of the Fair" mascot. They borrowed Nancy T for the event two years in a row and experienced the most successful publicity mascot animal in their history. The second year, the Fair painted cow footprints on all the roads leading to the Fair Symbol, chewing away in lush straw bedding. She was the delight of Orange County and an extremely productive sales representative, ensuring Fair success.

Nancy T remained part of the Herthel household for twelve years. Her daughter grew up to fill the gap when Nancy T passed on, but was never quite the same incredible family member. A few years later, in an effort to once again have a cow in the household, Sue and Doug traveled to Cal Poly, where they maintain a Jersey dairy herd for the students. At first the director tried to show Sue the cull herd, but that was a disappointment. Finally Sue was shown the operational herd, and she picked out the champion cow recognized as the best on the campus, but this was still not a Nancy T. We are all blessed when we find one animal who can grow a loving relationship and a place in our hearts and memories for all time.

Doug continues his successful practice for important horses from around the country and the Valley is fortunate and proud to have his clinic here. Clients should be aware, however, that his heart is still with a cow.

· 𝒯 ·

Country Day

IN THE seventies and eighties, a local mule breeder and trainer, George Chamberlain, ran a very successful mule operation on a ranch situated on Highway 154 at Foxen Canyon Road, just outside Los Olivos. Cars driving by would see a herd of handsome mules that were in training for packers and riding strings around the world. George was one of the main mule suppliers to the Maui Volcano rides that many have experienced. It was a unique way to make a living, but George was very good at it, and was also something of a character.

One Saturday morning he had arranged to entertain a number of his salty friends for a mule carriage ride about three

miles across country to our winery. There would be a barbecue picnic and of course a wine tasting, and then a return ride.
A number of us had trucked horses to the winery for the picnic and return, and our pal, Tom Lepley, had brought his good-
looking Hackney Horses and rig for the return ride. George's
two big rigs were a six-mule hitch with a large rubber-tired
wagon that he drove and a smaller four-mule wagon driven by
his top hand. Both rigs were full of friends out for a good time,
and all converged on the winery in good spirits. Three dozen
passengers, riders, and drivers rendezvoused and tied up at the
winery and proceeded to sample wine and barbecue.

The usual winery tour and tasting was put to full good
use and continued with total enthusiasm into the courtyard,
with the party making considerable inroads into the winery
inventory. We had a policy of not sampling too much of the
product on the premises, but all good rules are maintained by
the exceptions, and this was one of those sunny Saturdays in
the country made for exceptions. Here were old Valley friends
at ease and out for a good Western time, with horses, mules,
good food, and, of course, fabulous wine.

After lunch, we all saddled and hitched and began the parade across the hills toward the Chamberlain Ranch at the
corner of Foxen and Highway 154. George had the foresight to
bring a cooler, and we rested at the top of a hill and relived the
joys of wine tasting. The party continued, happily meandering down the hill by the County dump, where we would then
ride down Foxen Canyon Road.

Along the way, a dispute had risen between George with
his mules and Tom with his horses about who could outrun
the other. A six-up mule team and a Hackney Horse pair
would be an unusual race, and many joined in the discussion.
Soon money was out and pride was on the line. George loved
his mules and was not about to be undone by a pair of sleek
and fancy Hackneys. Tom thought he had the speed and class.

They agreed to take a run down the mile-plus of Foxen
Canyon Road from the County dump entrance to George's
gate, and tightened up for the contest. As the rigs approached
the designated start, I manifested my only glimmer of com-

mon sense that day. Bob Herdman Sr., in his eighties, was unsteadily perched on George's wagon with the other lunatics who were in for the race ride. I rode up to the side of the wagon. "Brother Bob," I said, "this might be a good one to sit out!"

"Think so, Brooks?" he replied, and climbed down to seek refuge in one of the motorized support vehicles that had now joined the parade.

Some of us cantered down the road ahead to warn off traffic. From my station, about three hundred yards down the road from the impromptu race start, I had a view of two teams boiling down the road, hell-bent for leather. Hats, bottles, and pieces of wagon flew into the wind as the passengers held on for life. Tom and George yelled and whipped as the frantic animals raced side-by-side down the road. I had warned off a carload of tourists who stood at their vehicle hypnotized by the crazy scene of a road taken over by the insane horsey locals. The flat-out Wild West show screamed past us and on down the road. I joined other riders in the gallop, and I still remember well my able ranch horse had a hard time keeping up with the six mule and two horse rigs. Neither gave way as the passengers hoped for survival.

It is recorded that the mules got the edge at the Chamberlain entrance, a hundred and fifty yards from State Highway 154, maybe because they thought they were going home. The drivers roared by the gate, heading for 154, hauling reins for their lives, to stop before the traffic. Somehow the sweaty, steaming, and heavy-breathing animals pulled up just short of a major highway wreck. Money changed hands, and the company settled down with oaths and relief. But the party wasn't over.

Someone mentioned that, on this day by pure chance, our pal Si Jenkins was opening his new store in Los Olivos, about a half-mile away. Si had run the best Western store in Santa Barbara for many years, and had decided to open a branch in the Valley, and this was the publicized Grand Opening. We all agreed that we should join his occasion and give the operation some more Western flavor. Soon the lathered and panting

horses, mules, and well-oiled pals with the wagons and support vehicles pulled up in front of the crowd of well-wishing
locals and tourists gathered for the event. Si, however much
he is a serious businessman, is also cowboy and a sport. He
found some cold beer and welcomed the Wild West show.
Toasts were drunk, and the party hit the afterburners.

I was riding Mandy Peake that day. She was a kind, talented stock horse mare who had the disposition to take care
of her less talented boss and get us through most horse-and-
rider situations unscathed. Sometimes her temperament and
calm were put to the test, and this was one of those occasions.

Inasmuch as this was a Western store, it seemed appropriate to me that both Mandy and I take a tour. We somehow got
through the door, and I rode Mandy around the aisles. With
great presence she ambled on an inspection of the merchandise quietly as though on a mountain path. After our horse-
and-rider browse through the merchandise, Si, who had held
his breath in the spirit of the occasion, welcomed me back
out into the street and Mandy relaxed, having saved her boss
once again.

In due course, things mellowed down, and George walked
his two rigs back home, Tom called for his trailer, and the
remaining company departed their separate ways. Kate was
kind enough to bring a trailer for horse and husband.

Mandy and I waited in the shade of a tree at Mattei's
Tavern, Mandy munching on the grass while I spoke grateful words of affection to her. I contemplated the extraordinary wealth of love and the dimensions of experience animals
bring to our lives.

· 𝓕 ·